Chemical Processing Aids in Papermaking: A Practical Guide

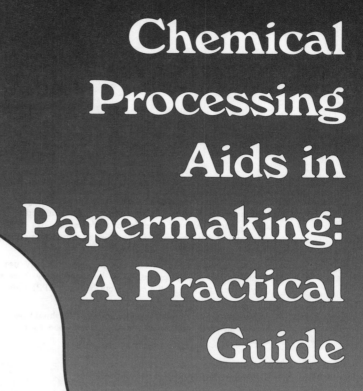

S0-ABA-845

Prepared by the Papermaking Additives Committee
of the Paper and Board Manufacture Division
Committee Assignment No. 5148

Philip M. Hoekstra, Chairman,
Chemical Processing Aids Subcommittee

Kevin J. Hipolit, Editor

TAPPI PRESS

1992

Vince Leary

1st Printing
Copyright © 1992 by

TAPPI PRESS
Technology Park/Atlanta
P.O. Box 105113
Atlanta, GA 30348-5113, U.S.A.

TAPPI Keywords: Agents; Aids; Alum; Deinking; Flocculants; Latexes; Microorganisms; Nonmicrobiological Deposits; Preservatives; Sizing Agents; Slimicides; Wet-strength Resins.

ISBN: 0-89852-256-0 • TAPPI PRESS Item Number 01 01 R192

Printed in the United States of America

Library of Congress Cataloging-in-Publication Data

Chemical processing aids in papermaking : a practical guide / prepared
  by the Papermaking Additives Committee of the Paper and Board
  Manufacture Division ; Kevin J. Hipolit, editor.
        p.   cm.
     "Committee assignment no. 5148."
     Includes bibliographical references and index.
     ISBN 0-89852-256-0
     1. Papermaking—Chemistry.   I. Hipolit, Kevin J., 1952-
II. Technical Association of the Pulp and Paper Industry.
Papermaking Additives Committee.
TS1120.C44   1992   92-2612
676'.2—dc20   CIP

TAPPI's ANTITRUST POLICY STATEMENT

TAPPI is a professional and scientific association organized to further the application of science, engineering, and technology in the pulp and paper, packaging and converting, and allied industries. Its aim is to promote research and education, and to arrange for the collection, dissemination, and interchange of technical concepts and information in fields of interest to its members. TAPPI is not intended to, and may not, play any role in the competitive decisions of its members or their employers, or in any way restrict competition among companies.

Through its seminars, short courses, technical conferences, and other activities, TAPPI brings together representatives of competitors in the pulp and paper industry. Although the subject matter of TAPPI activities is normally technical in nature, and although the purpose of these activities is principally educational and there is no intent to restrain competition in any manner, nevertheless the Board of Directors recognizes the possibility that the Association and its activities could be seen by some as an opportunity for anticompetitive conduct. For this reason, the Board has taken the opportunity, through this statement of policy, to make clear its unequivocal support for the policy of competition served by the antitrust laws and its uncompromising intent to comply strictly in all respects with those laws.

In addition to the Association's firm commitment to the principle of competition served by the antitrust laws, the penalties which may be imposed upon both the Association and its individual and corporate members involved in any violation of the antitrust laws are so severe that good business judgment demands that every effort be made to avoid any such violation. Certain violations of the Sherman Act, such as price-fixing, are felony crimes for which individuals may be imprisoned for up to three (3) years or fined up to $100,000, or both, and corporations can be fined up to $1 million for each offense. In addition, treble damage claims by private parties (including class actions) for antitrust violations are extremely expensive to litigate and can result in judgments of a magnitude which could destroy the Association and seriously affect the financial interests of its members.

It shall be the responsibility of every member of TAPPI to be guided by TAPPI's policy of strict compliance with the antitrust laws in all TAPPI activities. It shall be the special responsibility of committee chairmen, Association officers, and officers of Local Sections to ensure that this policy is known and adhered to in the course of activities pursued under their leadership.

To assist the TAPPI staff and all its officers, directors, committee chairmen, and Local Section officers in recognizing situations which may raise the appearance of an antitrust problem, the Board will as a matter of policy furnish to each of such persons the Association's General Rules of Antitrust Compliance. The Association will also make available general legal advice when questions arise as to the manner in which the antitrust laws may apply to the activities of TAPPI or any committee or Section thereof.

Antitrust compliance is the responsibility of every TAPPI member. Any violation of the TAPPI General Rules of antitrust compliance or this general policy will result in immediate suspension from membership in the Association and immediate removal from any Association office held by a member violating this policy.

General Rules of Antitrust Compliance

The following rules are applicable to all TAPPI activities and must be observed in all situations and under all circumstances without exception or qualification other than those noted below:

1. Neither TAPPI nor any committee, Section or activity of TAPPI shall be used for the purpose of bringing about or attempting to bring about any understanding or agreement, written or oral, formal or informal, express or implied, among competitors with regard to prices, terms or conditions of sale, distribution, volume of production, territories or customers.

2. No TAPPI activity or communication shall include discussion for any purpose or in any fashion of prices or pricing methods, production quotas or other limitations on either the timing or volume of production or sale, or allocation of territories or customers.

3. No TAPPI committee or Section shall undertake any activity which involves exchange or collection and dissemination among competitors of any information regarding prices or pricing methods.

4. No TAPPI committee or group should undertake the collection of individual firm cost data, or the dissemination of any compilation of such data, without prior approval of legal counsel provided by the Association.

5. No TAPPI activity should involve any discussion of costs, or any exchange of cost information, for the purpose or with the probable effect of:

a. increasing, maintaining or stabilizing prices; or,

b. reducing competition in the marketplace with respect to the range or quality of products or services offered.

6. No discussion of costs should be undertaken in connection with any TAPPI activity for the purpose or with the probable effect of promoting agreement among competing firms with respect to their selection of products for purchase, their choice of suppliers, or the prices they will pay for supplies.

7. Scientific papers published by TAPPI or presented in connection with TAPPI programs may refer to costs, provided such references are not accompanied by any suggestion, express or implied, to the effect that prices should be adjusted or maintained in order to reflect such costs. All papers containing cost information must be reviewed by the TAPPI legal counsel for possible antitrust implications prior to publication or presentation.

8. Authors of conference papers shall be informed of TAPPI's antitrust policy and the need to comply therewith in the preparation and presentation of their papers.

9. No TAPPI activity or communication shall include any discussion which might be construed as an attempt to prevent any person or business entity from gaining access to any market or customer for goods or services, or to prevent any business entity from obtaining a supply of goods or otherwise purchasing goods or services freely in the market.

10. No person shall be unreasonably excluded from participation in any TAPPI activity, committee or Section where such exclusion may impair such person's ability to compete effectively in the pulp and paper industry.

11. Neither TAPPI nor any committee or Section thereof shall make any effort to bring about the standardization of any product for the purpose or with the effect of preventing the manufacture or sale of any product not conforming to a specified standard.

12. No TAPPI activity or communication shall include any discussion which might be construed as an agreement or understanding to refrain from purchasing any raw material, equipment, services or other supplies from any supplier.

13. Committee chairmen shall prepare meeting agendas in advance and forward the agendas to TAPPI headquarters for review prior to their meetings. Minutes of such meetings shall not be distributed until they are reviewed for antitrust implications by TAPPI headquarters staff.

14. All members are expected to comply with these guidelines and TAPPI's antitrust policy in informal discussions at the site of a TAPPI meeting, but beyond the control of its chairman, as well as in formal TAPPI activities.

15. Any company which believes that it may be or has been unfairly placed at a competitive disadvantage as a result of a TAPPI activity should so notify the TAPPI member responsible for the activity, who in turn should immediately notify TAPPI headquarters. If its complaint is not resolved by the responsible TAPPI member, the company should so notify TAPPI headquarters directly. TAPPI headquarters and appropriate Section, division, or committee officers or chairpersons will then review and attempt to resolve the complaint. In time-critical situations, the company may contact TAPPI headquarters directly.

Statement of TAPPI Antitrust policy regarding submission of copies of correspondence to TAPPI headquarters

TAPPI headquarters needs to remain aware of what particular committees and sections of TAPPI are doing or planning to do in order to better assist those groups in achieving their objectives and to continue to supervise actively the antitrust compliance of TAPPI. The Board of Directors of TAPPI therefore has adopted this formal statement of TAPPI's policy which requires that persons corresponding or receiving correspondence on behalf of TAPPI provide copies of the type of correspondence outlined below to the appropriate liaison person at TAPPI headquarters.

For this policy TAPPI does not require copies of routine, written communications regarding arrangements for speakers, meetings, travel, dinner reservations and the like.

TAPPI headquarters does require that copies of correspondence of an important nature and of non-routine matters be supplied in a timely fashion to TAPPI headquarters personnel connected with the committee or section involved as shown below:

1. Plans regarding the activities of TAPPI committees or sections.

2. Communications with other TAPPI committees or sections.

3. Communications with persons or organizations outside TAPPI.

4. All written or recurring verbal complaints or criticisms of TAPPI activities.

All correspondence falling under the above-stated policy must be forwarded promptly to the appropriate TAPPI headquarters liaison person, preferably at the time of transmittal or receipt.

Acknowledgments

This publication is the combined dedicated effort of several contributing authors. TAPPI thanks these industry experts who took time from their regular jobs to assist with this work and to their supportive employers.

Contributing Authors

Philip M. Hoekstra - Buckman Laboratories, Inc.
Kevin J. Hipolit - PPG Industries, Inc.
Alan J. Schellhamer - Betz PaperChem, Inc.
Wallace E. Belgard - Betz PaperChem, Inc.
George S. Thomas - Betz PaperChem, Inc.
Linda Brooks Bunker - EZE Products, Inc.
Dr. William E. Smith - EZE Products, Inc.
Barbara H. Wortley - General Chemical
Robert E. Cates
Richard D. Harvey - Grain Processing Corp.
T. Small - Grain Processing Corp.
Alan J. Bauch - Pfizer, Inc.
Frederick Halverson - American Cyanamid Co.
James J. Scobbo, Sr. - Gencorp Polymer Products
J. R. Nelson - J. R. Nelson and Assoc.
John S. Conte - Quaker Chemical Corp.
Gregory W. Bender - Quaker Chemical Corp.
Tom W. Woodward - Betz PaperChem, Inc.

Preface

This monograph is intended to provide members of the paper industry with basic information on the use of additives commonly used in the papermaking process but not well understood by many. We hope this fills a need in the industry, and helps to make the papermaking process more efficient.

We take this opportunity to thank all those who, over several years of work on this project, put in much time and effort to prepare and revise the information presented here.

Philip M. Hoekstra, Chairman
Chemical Processing Aids Subcommittee
TAPPI Papermaking Additives Committee

Table of Contents

Chemical Processing Aids in Papermaking: A Practical Guide
© 1992 TAPPI PRESS

List of Tables and Figures

Chemical Processing Aids in Papermaking: A Practical Guide
© 1992 TAPPI PRESS

Introduction

Papermaking additives include a variety of materials added in small quantities at various stages in the papermaking process to produce a desired effect. The effect can be an improvement in the process, such as increasing the drainage rate, or a change in the physical character of the finished product, as in imparting a degree of sizing to the paper. The number and diversity of materials used as papermaking additives is quite large.

This publication deals with fourteen of the most commonly used classifications of additives. The chapters are intended as an introduction and general reference for the subject matter and, as such, are by no means complete. Many of the chapters are subjects for entire TAPPI monograms. Sources of additional information are included throughout the publication.

Each chapter provides a general description of the additive class, purpose, criteria for product selection, and general information regarding chemical and physical properties of the products. Also included is information on the safe handling and storage of additives.

The broad diversity of materials used as additives makes it essential for users to be familiarized with individual products and their safe and proper use. Suppliers are required by law to provide detailed information on the various physical and health hazards posed by a chemical using a standard format known as a Material Safety Data Sheet (MSDS). Product literature and labels also provide details regarding the safe handling of a particular chemical. This information should be read carefully and strictly adhered to. Observing common sense safety precautions and using proper protective equipment will minimize concerns related to handling chemicals.

Chemical additives are regulated to varying degrees by government agencies based on their toxicity and intended end use. The Environmental Protective Agency (EPA) is responsible for the implementation of various laws including the Toxic Substance Control Act (TSCA) and the Resource Conservation and Recovery Act (RCRA). All slimicides in the United States are registered through the EPA which states all acceptable uses of the product. The Food and Drug Administration (FDA) regulates chemical applications in the manufacture of paper and paperboard where the intended use is for direct contact with food. These restrictions are published in the Code of Federal Regulations under Titles 40 and 21, respectively. Finally, the Occupational Safety and Health Administration (OSHA) governs the use of all hazardous chemicals in the workplace under the Hazardous Communication Standard 29 CFR 1910.1200. The Employee's Right-to-Know Law requires employers to provide information on the hazards of chemicals in the workplace. Information on specific additives can be found in the individual chapters.

Kevin J. Hipolit
Research Associate
PPG Industries, Inc.

Chapter 1

Slimicides, Preservatives, and Other Microorganism Control Agents

by Philip M. Hoekstra

Introduction

It seems unlikely that a huge modern paper machine could be shut down by organisms which are difficult to see even with a high-quality microscope. It doesn't seem possible that these same microscopic creatures can cause holes and odors that reduce the quality of the product from a high-tech paper mill. But the efficient operation of a computer-controlled, nine-meter-wide marvel of modern papermaking technology depends on keeping the growth of these seemingly insignificant microscopic organisms under control.

Almost all paper machines are periodically affected with problems caused by microscopic organisms, problems we usually refer to as "slime." When slime appears in a paper machine system, problems begin. The news that the machine is covered with slime can cause panic in the mill. A normally civilized and calm machine superintendent can be turned into a dangerous creature to avoid at any cost.

Slime-related problems are a huge economic drain. The Institute of Paper Chemistry estimates that slime-related problems cost the United States paper industry upwards of $100 million. Other sources estimate that cost would be ten times higher without slime control systems and chemicals.

Controlling slime is a topic that is not well-understood. In this chapter, suggestions will be presented on how best to control slime problems. Answers to practical questions regarding slime will be discussed. What conditions on the machine encourage slime growth? How much slime control chemical should be added? How does the use of recycled fiber affect slime growth? How important are boil-outs?

Problems Caused by Slime

Controlling slime can be costly, especially on a modern alkaline fine paper machine. However, the costs of not controlling slime can be much higher. Often the greatest cost of slime-related problems is lost production. On a modern high-speed paper machine, an hour of down time may cost more than $20,000 in lost production. On some high-speed linerboard machines, one break may cost more than that. And if a machine does not reach its alloted production schedule, it can have a negative impact on careers of mill personnel. Table 1 lists several problems caused by slime on a paper machine.

There is no doubt that, in order to efficiently produce quality product at its highest production rate, a modern paper machine needs an effective slime control program. Before outlining specifics of how best to control slime, a clear understanding of the nature of slime is in order.

What is Slime?

Obviously slime gets its name from the fact that is it often a wet, sticky, gelatinous, generally unpleasant material. However, slime can occur with a variety of characteristics. It can be rubbery, stringy, granular, or pasty. Slime with an appearance of tapioca grows in both unbleached kraft linerboard and alkaline fine paper machines.

Since slime collects solids in the furnish, it can take on any color. It is not uncommon for the dreaded "pink slime" to occur in a machine. This deposit is caused by certain microscopic organisms and has a typical orange, red, or pink color. Pink slime has been seen on fine paper machines running at acid conditions, and also on some fine paper machines that have converted to neutral or alkaline conditions.

Although slime has many different appearances, not every slimy deposit is indeed slime. True slime is always produced by the tiny microorganisms mentioned previously. Verifying the presence of slime requires microscopic observation of these microorganisms.

The microscopic organisms causing slime often grow in stringy forms called filaments. These microscopic filaments form a type of net that traps fines, fillers, dirt, and water which make up a slimy mixture. Also, many of the bacteria growing in systems like paper machines actually manufacture slime as a means of protection.

Figures 1 and 2 illustrate masses of slime on paper machine surfaces. In each case, bacteria have attached to a surface, begun to grow, and built up a slime deposit.

Table 1. Problems Caused by Microbiological Growth in a Paper Mill

Loss of production	Breaks, unscheduled boilouts, and washups
Reduction in quality	Slime spots, holes, odors in product
Microbiological corrosion	Damage to metals, including stainless steel
Loss of additives	Microorganisms can grow in any of the following and plug lines, cause odor, reduce the effect of the additive, or cause more slime growth on the machine: coatings, broke, fillers, wet lap, alum, retention aid, size, starch (cooked or uncooked)
Production of dangerous gases	Hydrogen, hydrogen sulfide, or methane can be produced by bacteria, creating the potential for explosions
Off-spec and rejected product	Cost of segregating and repulping broke, cost of customer claims, loss of customers

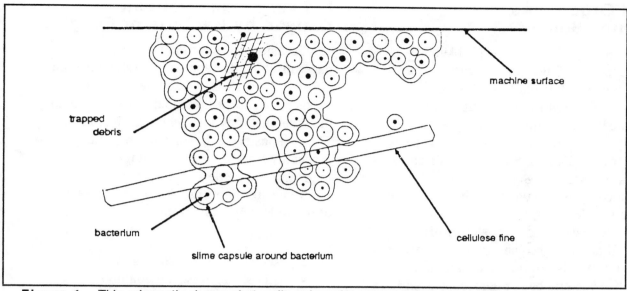

Figure 1. This schematic shows what a slime deposit on a paper machine might look like. Many bacteria attach to a surface and grow there. The bacteria shown here excrete a layer of slime to help them attach and protect themselves. The slime traps debris such as fines and fillers flowing past. If the growth is not controlled, the mass of bacteria, slime, and debris continues to build. Problems in production and quality eventually begin.

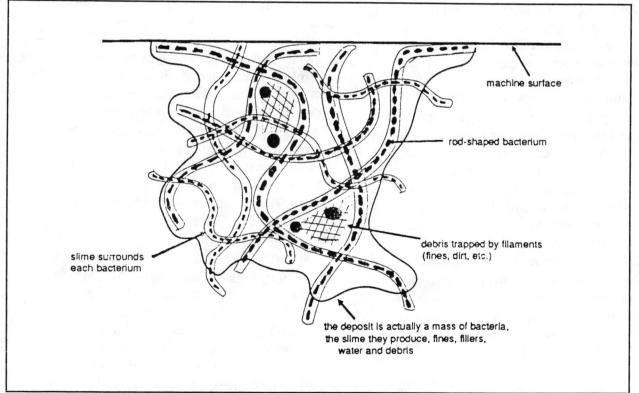

Figure 2. This schematic shows a deposit caused by another common form of bacteria. Many types of bacteria grow in long stringy filaments which are attached to surfaces in the paper machine. The slime produced by the bacteria holds the filaments together and this mass of slimy filaments traps water and debris. If the growth is not controlled, the mass of bacteria, slime, and debris continues to build. Problems in production or quality eventually begin.

What Organisms Produce Slime?

Organisms causing slime problems are more accurately called microorganisms due to their small size. It takes 15,000 typical bacteria placed end-to-end to stretch one inch. In most cases, the culprits causing a slime deposit are bacteria. Several thousand species of bacteria are known. Some cause diseases, others are beneficial. Some species of bacteria will produce odors, others can corrode stainless steel.

In other cases, usually at lower pH levels, slime might be caused by the growth of fungi or molds. Fungi are usually much larger than bacteria but a microscope is still required to see the individual cells. Although fungi do not excrete slime, they are made up of stringy filaments which trap fines and debris, thus forming slime deposits.

Occasionally other microscopic creatures are found in a paper mill system, but these are seldom the primary cause of a problem. For example, yeast often are in deposits, but the only place yeast are likely to be a problem is when they grow in a starch slurry and cause fermentation. Algae can occasionally cause a paper defect, usually a green spot in a sheet, and that is usually due to contaminated incoming fresh water, not the paper machine.

All of these creatures are very small, but their numbers are very large. One gallon of water from a clean river might easily contain millions of bacteria. Multiplied by the number of gallons a mill uses in a day, a huge number of microorganisms come into the mill. In addition, some of these organisms can multiply themselves very quickly. When conditions are right, numbers of organisms can double every twenty minutes.

Is Every Slimy Deposit Caused by Bacteria or Fungi?

Occasionally, deposits can form in a paper machine that appear to be slime but are not actually caused by bacteria or fungi. Such deposits can break loose and cause defects in the sheet that look identical to a slime hole. These deposits can be difficult to identify, especially since bacteria and fungi can begin to grow after the deposit forms.

Slime-like deposits can be caused by any of the following additives:

1. Alum

2. Recycled paper materials such as adhesives, inks, and coatings

3. Retention aids

4. Starch

Alum has a limited solubility in water. Particularly when the alum is overused, or the pH of the system is near neutral, alum deposits can occur. At times this can be localized to one spot. For example, if the shower water in the headbox is not pH-controlled, alum can deposit where the furnish comes in contact with the shower water. This deposit often has the appearance of a slimy gel. If the rosin/alum ratio is not correct, deposits made up of size or alum or both often result. This can happen when the pH control on the machine is not very precise. The pH probes should be cleaned and calibrated frequently to make certain the pH reading is accurate.

Materials from recycled paper (adhesives, inks, and coatings) can cause deposits which are often called stickies. If coated broke is recycled in the mill, the "white pitch" that often deposits on the machine can have a slime-like appearance. Occasionally, certain types of additives used to control foam might be involved in these deposits. When these stickies deposits break loose, the resulting holes or breaktails can look just like holes caused by slime.

Some slime-like deposits are related to retention aids. Deposits looking very much like slime on the wet end of a machine, when observed with a microscope, actually show very little microbiological growth. In some cases these deposits may be difficult to diagnose, since bacteria and fungi might begin to grow very quickly in such a deposit. In these cases, perhaps the only way to prove the source of the deposit is to temporarily turn up the slimicide. If the deposit does not go away, there must be some other cause.

Starch is another additive sometimes involved in slime-like deposits. If the starch is not properly prepared, or not retained well in the sheet, it may deposit on the surfaces. In this case, bacteria almost immediately begin to grow in the deposit, and it looks like a slime deposit upon inspection. However, the best solution is to solve the starch problem, not add slimicide.

Deposit formation can occasionally evolve into a chicken-and-egg situation when trying to determine whether the slime came first, or some other material deposited, creating a place for slime to grow.

Factors Affecting Microorganism Growth

A first step in understanding and controlling slime is learning what conditions in the paper machine affect the growth of microorganisms. Like the other living things on earth, bacteria and fungi need food, water, and certain other essentials.

Water

Water is an important factor, since water makes up the total environment of the organisms discussed here. Water carries all the nutrients, controls the temperature and all other conditions in their environment. Water

transports the slimicides used to control these organisms.

Food

The food enabling bacteria and fungi to grow is carried by water in the system. Therefore, the quality of water used and the additives in the water affect bacteria growth. Adding starch, for example, into the system can cause a sudden increase in slime.

Temperature

Temperature is another important factor in slime control. Just as the survival of any animal or plant is determined by the temperature of its environment, the survival and growth of microorganisms depends on the temperature of the paper machine. Figure 3 illustrates the approximate effect of wet-end temperature on slime. As an example, if the wet-end temperature is normally around 50°F (10°C), and if it drops to 40°F (4°C), there is likely to be an increase in slime, since bacteria and fungi grow better at this more moderate temperature.

pH

The acidity and alkalinity of the system has a significant effect on slime problems. Most fungi grow better in acid conditions while most bacteria prefer neutral conditions. If the paper machine has operated at acid conditions (for example, with rosin size), and then begins producing paper at neutral or alkaline conditions (for example, making grades with calcium carbonate filler), slime problems are likely to be much more severe. This is illustrated in Figure 4.

Oxygen

Most microorganisms can thrive in the presence of molecular oxygen, however, anaerobic organisms cannot tolerate oxygen in that form. The stock in most of the paper

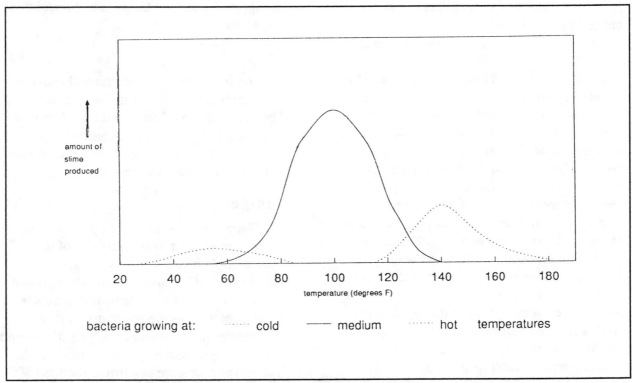

Figure 3. Effect of Temperature on Growth of Bacteria
As the temperature changes in a paper machine, the amount of slime produced can change dramatically. Slime problems are usually most severe at temperatures of 90-110°F.

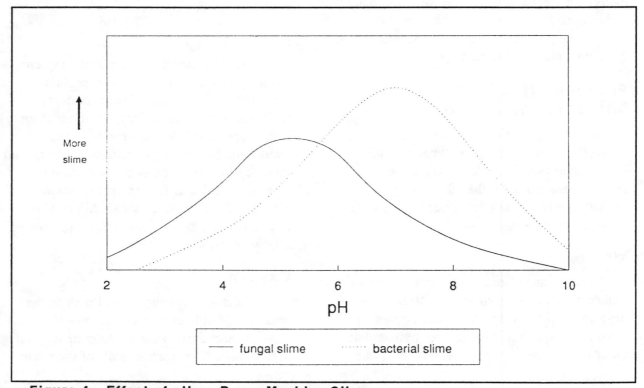

Figure 4. Effect of pH on Paper Machine Slime
pH can have an important effect on slime problems in a paper machine. Problems tend to be much worse on the alkaline side (around pH 7) compared to acid papermaking (pH 4-5).

machine is well-aerated (and so contains much oxygen). Oxygen is not present in the areas under deposits or in chests that are not agitated well. It is in these areas where anaerobic bacteria can thrive. Anaerobic bacteria cause the odors known as sour stock which are occasionally produced in the chests or in the paper product. Other types of anaerobic bacteria can cause corrosion and produce gases such as methane, hydrogen, or hydrogen sulfide which are poisonous and have caused explosions in some mills.

Environmental Factors in Slime Growth

Temperature

Most slime growth occurs at 100-110°F (38-43°C) with no fungal growth above 122°F (50°C) and little slime growth above 140°F (60°C).

pH

Most slime growth occurs in neutral conditions with less growth in acid conditions and little or no growth in neutral conditions.

Oxygen

Oxygen-free conditions can exist underneath slime deposits and result in additional problems. Corrosion, odors, and poisonous gases occur in oxygen-free environments.

Basics of Controlling Slime

It is generally agreed that an effective slime control program is a necessary part of operating today's high-technology paper machines. How best to control slime is a complex matter.

Slime control is not simply adding chemicals to the system to kill trouble-causing organisms. Costs can usually be reduced with an effective slime control system in place. Table 2 details common uses and points of addition of slimicides.

Reduce Contamination

A very important part of controlling slime involves minimizing trouble-causing organisms entering the paper machine. This involves keeping the whole system as clean as possible; from the point raw materials

Table 2. Paper Mill Uses of Microbicides

Points of Addition	Intended Use
To paper machine stock or white water	Control growth of slime on machine surfaces
To broke	Prevent microbial growth and odor
To savealls	Prevent growth of slime which reduces efficiency
To starch slurry or paste	Prevent pH drop and odor
To clay or other fillers	Prevent odor
To coating	Prevent slime and odor
To high-density storage	Prevent odor
To incoming fresh water	Reduce slime growth in areas such as fresh water showers
To various other additives, such as dyes, retention aid, etc.	Prevent microbial breakdown and slime in screens
To alum	Prevent growth of fungi

enter the system through the machine to the reel and converting. Mills maintaining clean systems will have fewer slime problems.

Washups are an integral part of the process of reducing contamination. When the machine is down for a break or boil-out, or on a grade change, water hoses should be used to spray areas prone to a buildup of stock or deposits. Controlling other types of deposits on the machine is very important. If pitch, stickies, scale, fiber, or filler build up in the system, ideal places are created for bacteria and fungi to grow. It may be necessary to coordinate the use of dispersants with slimicides.

Probably one of the most important parts of slime control on some machines is a proper boil-out. Some machines may run indefinitely without a boil-out; others may need to boil out every two weeks. The boil-out frequency depends on many of the same variables any part of the slime control program depends on.

In a boil-out, hot water is circulated throughout the machine and white water circuit. For maximum effect, the temperature should be at least 140°F (60°C). Addition of caustic soda usually makes the boil-out much more effective, since many deposits are dissolved in a strong alkaline solution. A pH of 12 should be maintained throughout the boil-out. In many cases, the use of a dispersant can result in a major improvement by helping to penetrate and disperse slime and other deposits.

The frequency of boil-outs is often closely related to slimicide use. Frequent boil-outs require less slimicide to control slime on the machine. However, if a slime control program is running well, and the running time between boil-outs is extended from three to five weeks, slime problems may surface. To extend time between boil-outs, it may be necessary to increase the slimicide usage.

Slime Control Chemicals

Another important element of slime control is the choice of chemical product used to control slime growth. A variety of products exist, most are commonly known as slimicides or microbicides, and there are numerous uses for these products. The information provided by suppliers should be read carefully to make certain the use of the product is as safe and effective as possible.

Every slimicide used in the United States should be registered with the Environmental Protection Agency (EPA), which supervises testing and authorizes the label for each product. The label is a legal document which lists all acceptable uses for the product.

Slimicide manufacturers take products through long periods of testing to determine effectiveness, evaluate toxicity, and numerous other factors in order to get the product registered by the EPA. Suppliers must satisfy both themselves and the EPA that the product is safe and effective. Every product registered by the EPA has gone through years of study.

It is very important that these chemical additives, like any material in the paper mill, be transported, handled, and stored as safely as possible. This means knowing about any possible hazards related to a product and using the proper safety equipment.

Safe Use of Slimicides

Slimicides in the United States have gone through a lengthy registration process with the EPA. These materials are safe to use, provided they are used according to the Material Safety Data Sheets (MSDS) and product labels in conjunction with common-sense safety precautions. The potential for problems is minimized when these products are handled correctly.

The first rule of safety for using slimicides is to read the MSDS and product labels. These will identify any hazardous components and any health hazards associated with the particular microbicide.

The second rule of safety is to use proper protective equipment. Any chances for human contact with the slimicide must be minimized. Most microbicides can cause skin irritation and eye damage when there is direct contact. The importance of personal protective equipment cannot be overemphasized. Chemical-resistant gloves, safety glasses with side shields, and chemical-resistant boots should be available to those handling slimicides. Adequate ventilation is important in areas where slimicides are stored, and in any area where there might be misting.

The slimicide supplier should provide health and first-aid information, and information on medical treatment of any problems that arise. In general, proper first aid for eye contact means immediately flushing eyes with water. There should be eye-wash stations near areas where contact with the product is possible, and these stations should be well-marked. The first two minutes are crucial to minimize damage to eye tissue. The employee should wash eyes for a minimum of 15 minutes, then seek medical attention. If a microbicide comes in contact with the skin, the initial treatment is to flush with plenty of water. All contaminated clothing should be removed. Seek medical attention if necessary.

Location of Slimicide Containers and Pumps

1. Choose a location with less chance of human contact.

2. Carefully label all containers and tubing carrying slimicides.

3. Use quality pumps, timers, and tubing.

4. Do not add slimicide by hand with containers.

5. Make certain the label is readable and the MSDS is available nearby.

6. Surround the slimicide storage area with a dike to prevent product from spreading in the event of a leak or spill.

7. Consider using semi-bulk or bulk containers if the quantity used is significant. Use of larger containers minimizes contact with product involved in moving and disposing of drums.

8. Tubing delivering slimicide to the machine should be encased in a protective outer shell made of metal or PVC if tubing passes over an area where people work or congregate; if there is any possibility of pressure building up in the line feeding slimicide; if there is a possibility of all or part of the line getting so cold that the slimicide might freeze, resulting in a pressure buildup.

Pumps

1. Interlock electrical power to the pump with the machine system to prevent pumping slimicide when the machine is down.

2. Frequently inspect, maintain, and clean pumps, containers and the surrounding area.

3. Pumping equipment should be fitted with pressure-relief valves. If part of the line is obstructed by frozen slimicide, pressure could burst the line unless a pressure-relief valve is in place. A common set-up uses a pressure-relief valve set for approximately 70 psi along with

tubing rated to a minimum of 120 psi. Operators must know the pressure of the part of the system where the slimicide is being fed.

Protective Equipment and Safety

1. Locate showers and eye washes near slimicide containers and pumps.

2. Have face shields, gloves, and other appropriate protective equipment available near slimicide containers, pumps, and lines.

3. NEVER eat or smoke when working with microbicides.

4. NEVER wipe eyes with a cloth, glove, or hand that has been in contact with a microbicide.

5. ALWAYS have a healthy respect for any chemical product.

6. ALWAYS protect eyes by using a face shield when handling slimicides.

7. ALWAYS wash hands before eating.

8. ALWAYS read the MSDS before working with a new product.

9. ALWAYS know which products are compatible and which are incompatible for use in storage or pumping of the product.

Government Regulation of Slimicides

In most countries, the use of slimicides and similar products is regulated to some degree. The maze of regulations controlling the use of products such as slimicides continues to expand. In addition to the several federal agencies involved, most states have regulations as well. This section outlines the main federal agencies controlling slimicides in the United States.

The role of the EPA is to protect the environment. Because the EPA is the lead agency in the implementation of the Toxic Substances Control Act (TSCA) and the Federal Insecticide, Fungicide, and Rodentcize Act (FIFRA), this role includes controlling the use of chemical products and protecting users of pesticides. Part of the duty of the EPA is to control both what pesticides are registered and how those registered products are labeled. This makes the label on a slimicide container the most important document associated with that product. No pesticide may be sold without a label approved by the EPA. It is clearly illegal to use a pesticide for a use not described on the label, or to use that product at levels not permitted on the label. The label for every pesticide used in the United States must contain the following statement: "It is a violation of Federal law to use this product in a manner inconsistent with its labeling." A violation carries penalties of one year in prison and up to $25,000 per day per violation in fines.

The Occupational Safety and Health Administration (OSHA) governs the use of all hazardous chemicals in the workplace in order to protect workers under the Workers Right-To-Know Act. OSHA regulations govern MSDS and right-to-know compliant labeling for all products except pesticides, which are handled by the EPA. OSHA also has a carcinogen policy requiring labeling, warning, and measurement of any carcinogens in a workplace. OSHA is responsible for determining safe levels for air contaminants. Under right-to-know laws, OSHA requires mills to train all people involved with hazardous materials. If a chemical product is in the mill, an MSDS must be available to all employees in order to communicate any hazards associated with a product and allow workers to be aware of those hazards.

The FDA governs anything that is or can become a component of food or drugs, which by definition includes packaging for those materials. Additives and all other components of the sheet, if these materials can migrate from the paper to the food, are considered "indirect food additives." A paper mill must be concerned with the FDA only if some of its products come in contact with foods or drugs. If a certain additive is used, it requires FDA approval only if it is shown to become a component of the paper or board.

The TSCA is enforced by the EPA and applies to all chemicals that are not pesticides or food additives. All chemical components of any product manufactured or imported into the United States must be on an inventory of known materials, or that product cannot be used in the United States The end user doesn't need anything more than a letter from his supplier to comply with TSCA.

The Resource Conservation and Recovery Act (RCRA) is also enforced by the EPA, and governs the disposal of hazardous wastes.

The Comprehensive Environmental Response, Cleanup, and Liability Act of 1980 (CERCLA or Superfund) was amended by the Superfund Amendments and Re-authorization Act of 1986 (SARA). SARA Title III deals with at least four areas of interest to paper mills relating to slimicides and other chemical products including requirements dealing with an Extremely Hazardous Substance list and a Toxic Chemical list. Also included is a section dealing with community right-to-know legislation. These laws require notification of local, state, and federal agencies of the presence of certain substances at a facility, reporting accidental releases, reporting inventories, and reporting normal releases.

Common Questions Regarding Slimicide Use

This section will address the following questions, among others, regarding the use of slimicides:

How much slimicide should be used? Which slimicide should be used? Where should the slimicide be added to the system? Should the slimicide be added continuously? Should slimicides be alternated? Should two or more products be used? How often should the system be boiled out? How is slimicide effectiveness determined? Will the slimicide affect any any properties of the paper product?

For many of these questions there are no straightforward answers. The products used, the cost per ton, and many other factors vary widely, especially with the grade of paper produced.

What is the Problem?

One should first consider the nature of the problem. Ask the following questions: Is this slime just a cosmetic problem or is the deposit of major importance? Will customers be lost if the problem recurs? How much production time in man-hours will be lost if the slime-related problem gets out of hand? Mill operators should step back and evaluate the costs and benefits of the slime control program. Other questions can then be addressed.

How Much Slimicide Should Be Used?

The answer to this question is obvious: Use enough slimicide to prevent any quality or production problems on the machine.

The cost of a slime control program will vary from mill to mill. A machine running very hot, above 140°F (60°C), or a machine running heavyweight board (unlikely to have holes or breaks from slime), or a machine making many grade changes with washups between changes, may not need any slimicide. On the other hand, certain fine paper machines running at alkaline pH with a number of additives (size, starch, dyes, retention aids, etc.) may need a slime control program 10 times as costly as those in other mills.

The concentration of actives varies from one slimicide to another, thus affecting the amount of product used.

There are ways, however, to gauge optimum levels of slimicide. For example, if the slimicide feed is accidentally cut off, how soon do signs of slime growth appear? That is, how soon do machine surfaces begin to get slippery? If slime problems appear within a day or two, operations are close to the minimum usage on that slime control program. Another method used by some paper mills to gauge optimum levels of slimicide involves running past the normally scheduled boil-out and noticing how soon slime problems appear.

To determine whether too much slimicide is being used, cut back on the use rate and observe any changes. Gradual rather than dramatic reductions will allow more accurate assessments. Cut back 10 percent and let the machine run for a couple of weeks. Use a little more than the minimum amount of slimicide to provide insurance in the event that a change in the system suddenly creates favorable conditions for the growth of slime.

When comparing one paper machine to another, there is not a straight answer on the question of how much slimicide is enough. There are several important factors in slime control which vary greatly from mill to mill. It is possible for two machines side-by-side in the same mill to have different slime problems. The following factors can be important in determining the amount of slimicide needed:

1. **Type of fiber used:** Fully bleached chemical pulp will contain fewer microorganisms and much less food for microorganisms than mixed waste.

2. **Types of additives used:** Fillers, sizes, and especially starches provide food for slime growth.

3. **Quality of water used:** The raw water supply for a mill transports slime-producing organisms as well as food for their growth. The quality of this water may change from one time of year to another. For example, during spring runoff, slime problems increase in many mills.

4. **Temperature:** System temperature is very important in controlling slime growth. In machines running at about 140°F (60°C), there are few major slime problems. In some systems, machine temperatures vary with the season of the year. Some mills find serious fungal slime problems occur.with cooler winter temperatures.

5. **pH:** Fine paper machines that have changed from rosin/alum sizing systems to running alkaline size and alkaline pH have usually found the cost of slime control rise significantly.

The use of slimicides in satellite systems can be a very important part of keeping a paper machine running efficiently. The uses of slimicides should be optimized to reduce program costs. However, mills should be conservative in reductions as too great a reduction of preservative levels in starch or coating can cause spoilage, a problem which can cost much more than might have been saved by making an incremental reduction in slimicide usage.

It should be reemphasized here that it is illegal to use slimicides at levels higher than those recommended on the label.

The slimicide supplier should carefully monitor the use of these products to make certain the program runs efficiently. The supplier and the mill should work together to maintain a satisfactory supply of product, maintain proper operation of feed equipment, and communicate any system changes warranting changes in the slime control program.

Which Slimicide Should Be Used?

There are many slimicides used by the paper industry, each working well in many situations. A recent computer search of products registered by the EPA for use in the U. S. paper industry found more than 90 active ingredients in use. Some of these are no longer widely used, and so the list of commonly used materials is much shorter. Although it is not feasible to cover all available slimicides in this discussion, most available products can be grouped into general categories featuring advantages and disadvantages of products in each category. Table 3 lists potential advantages and disadvantages of slimicides in six categories. These are not absolute statements but are generally true for the products in each category. Also, many products are combinations of different types of active ingredients. A slimicide supplier should be able to provide information and a history on the use of the product recommended.

In some cases, preliminary testing can indicate a product's effectiveness within a system. For example, slimicide testing for the wet end of a paper machine might include kill studies involving treatment of samples of white water and slime with different slimicides, and allowing incubation under conditions similar to those of the paper machine. Periodic samples are analyzed to determine bacteria or fungi levels in the mixture after a specified waiting period. Testing for an effective product to use in a starch or coating might involve treating the material with various slimicides and observing how well the product controls the pH over a period of time. In the case of starch, bacteria and yeasts produce acids which make the starch more acid as time goes by.

Following are questions a slimicide supplier might consider before recommending a certain product:

1. **Is the slime problem primarily due to bacteria or fungi?** Many products can be effective in the control of bacteria. However, many of these are not that effective if the slime is due to fungi.

2. **What is the pH and temperature in the system?** Many products can be effective in controlling bacteria in a paper machine running at acid pH (for example, with alum/rosin sizing), but not many products are effective in alkaline systems (for example, with machines using calcium carbonate fillers).

3. **Is the paper or board to be used for packaging food products?** In the United States, all slimicides must be registered by the EPA. Almost all of those registered are also listed as allowed by the FDA in the manufacture of paper or board for food packaging. However, only a few of these are allowed for use in dry end applications in paper destined for food uses. So, if paper or board products are to be used for some food-contact application, and a

product is needed to treat, for example, the size press starch or coating, there are not many products allowed for such a use. (See the earlier section on FDA allowances for slimicides used in paper manufacture.)

4. Have there been past problems of odor or lacrimation due to slimicide? Certain slimicides are more likely to cause odors or irritation of eyes and nose. This is not a common problem, but can be a problem if the slimicide is used in

Table 3. Advantages and Disadvantages of Major Classes of Slimicides

Class of Product	Advantages	Disadvantages
Organobromines	Good bactericides Widely used	Possible effect on size Possible foam Some have odor problems Not effective on fungi Dissolved in solvents
Organosulfurs	Water-based Good bactericides Some are fair fungicides Widely used for many years	Very high pH Occasional odor problems Possibly detrimental effect on brightness in some machines
Cationics	Effective in some situations	May not work well in systems with high levels of solids
Isothiazolinones	Good bactericides Water-based	Less effective on fungi Some problems with irritation and skin sensitization
Thiocyanates	Broad spectrum Control of fungi and bacteria	Some problems with irritation and skin sensitization Not stable at pH levels above 8
Oxidizers	Widely used for water treatment Low cost-per-pound	Must add continuously Can damage machine fabrics Not effective on all microorganisms

shower water. Sometimes odor will be a problem in areas with poor ventilation.

5. **Is paper brightness very important?** Certain types of slimicides have been known to reduce the brightness in some high-brightness papers. This is not common but should be considered when choosing a slimicide.

6. **Can the supplier provide the necessary service and technical information?** Regarding the use of all specialty chemicals in a paper mill, but especially in the area of slimicides, communication with the slimicide supplier is very important. The level of know-how and service the supplier can provide should be considered.

The choice of which slimicides to use should be made jointly by the papermaker and the supplier with frequent communication throughout selection and testing. The supplier knows the technology and history behind the products to be used, and the papermaker knows the needs of his particular mill.

Where Should Slimicide Be Added to the System?

To answer this correctly, the nature and location of the slime-related problem must have been identified. The slimicide should be added to the system at the point where it will control the major problems.

The most important place to keep free of troublesome deposits is the wet end of the machine, the primary white water circuit. Slimicides are not intended to treat the stock in the machine as much as the machine surfaces in an attempt to keep those surfaces clean. There might be billions of bacteria circulating in the white water, but holes and breaks due to slime only occur when these build up on the surfaces of the paper machine. The most important goal in a slime control program is to keep slime from building up on the machine surfaces. Keeping the primary circuit clean almost certainly requires adding slimicide at the fan pump, to the tray water, or to the machine chest. It may be useful to experiment with different addition points in this primary circuit.

A successful strategy in many slime control programs is to treat, in addition to the wet end, any major sources of contamination to the machine. In other words, treat the part of the system where slime-formers are originating. Often, this is the broke system. Slime may develop from one of the additives, such as starch. In some cases, it might not be possible to control slime unless the raw water coming to the machine system is treated to reduce contamination.

With all slimicides, it is important to achieve proper mixing of the product into the system for maximum effect. While some products dissolve easily in water and others do not, mixing is important in every case.

It is also important to know the flows in the system in order to ensure the slimicide will get to where it can control the problem with maximum effect and minimum cost. For example, if a slime deposit existed above the waterline in a headbox, treating the stock entering the headbox might not eliminate the problem. In this case, the slimicide should probably be added to the shower water in the headbox.

Additional information on the system is necessary to determine an effective addition point. In one case, a slimicide was added to the machine chest in a mill making linerboard. Depending on upsets in the pulp mill, the pH of the stock at the machine chest might be in the range of 5-9. At the high pH, this particular slimicide (active ingredient methylenebis thiocyanate) was destroyed very quickly, with slime problems resulting. In this system the machine chest was not an acceptable addition point for this slimicide.

Should Slimicides Be Added Continuously?

Having decided which slimicide to use and where to add it, the choice of adding it continuously or periodically must be made. In most cases, wet-end slimicides are not added continuously, but intermittently.

If a machine surface receives a strong treatment for one hour, it will take some time before the microorganisms begin to grow again and reproduce. By that time slimicide is readded into the system. Often the treatment is done two to three times per day. It turns out that intermittent treatment works very well for controlling microbiological growth. By treating periodically, less slimicide is needed to control the slime growth.

There are other cases where continuous treatment is necessary. For example, to preserve starches or coating, slimicide must be continuously present in the starch to prevent excessive microbiological growth.

Should Slimicides Be Alternated Within the System?

Experience with slimicides over the years has shown it is almost never necessary to alternate slimicides. If a slimicide has worked well for a number of years within a system, a sudden increase in slime is not likely to be precipitated from organisms becoming resistant to the slimicide. There is the slight possibility that some new type of bacteria or fungi have entered the system. It is more likely that some temporary or permanent change in the conditions of the system (temperature, pH, water quality, etc.) has affected the slimicide.

Should Two or More Slimicides Be Used?

It is imperative to keep the wet end of the machine free from slime problems. In addition, on many machines it may be necessary to treat satellite systems (savealls, broke, starches, retention aids, alum, etc.) to prevent serious contamination of the wet end; to treat the water entering the system; or to treat size presses or coaters. So it is very common for an effective slime control program to utilize a variety of slimicides in several different parts of the system.

Will Slimicide Affect Paper Properties?

Slimicides infrequently affect some properties of the paper produced. Sizing, color, or brightness may be affected. Preliminary lab testing can indicate whether one of these properties would be affected.

The obvious solution to such a problem is to change slimicides. However, it may be equally acceptable to eliminate the problem by changing the addition point. For example, if there is a problem with a reduction in sizing, this might be solved by adding the slimicide after addition of the sizing agent instead of prior to its addition. In other cases, changing from intermittent addition of slimicide, to a continuous addition of a smaller amount of slimicide might yield desired results.

How Often Should Paper Machine Boil-Out Occur?

Boil-out frequency varies widely in the industry, from weekly or biweekly in some mills to never in others. In some cases, boil-out is not important, while in others it is not possible to maintain effective slime control without an effective boil-out program.

Boil-outs tend to be more important in complex systems using a number of additives (fillers, size, dyes, etc.). Boil-outs are also more important in mills that do not change grades often, and in mills with more serious slime problems.

Why boil out the paper machine? First, properly designed boil-outs will remove many types of deposits from the machine. In addition to slime, the most common deposits come from stickies, pitch, size, and alum. These deposits provide protected areas where slime-formers can grow unaffected by slimicides in the system.

Second, a good boil-out reduces the amount of slime-forming organisms on the machine. Reducing contamination is a primary factor in controlling slime on a machine. An effective boil-out is the best way to reduce contamination.

Determining Slimicide Effectiveness

There are a variety of ways to monitor effects of a slimicide. The bottom line is this: is it controlling slime problems or not?

There are a variety of ways to monitor slime control programs, but the best method involves monitoring the surfaces on the paper machine and locating those which are the first to become slippery. This indicates that slime growth is about to become a problem and a boil-out may be necessary. Operators who know the paper machine well can tell by daily inspection of machine surfaces whether a slime problem is under control or about to become a problem. Trouble spots should be identified early and monitored closely for optimum results.

Much time, money, and effort is wasted in the industry using plate counts to monitor slime programs. Plate counts are usually done with the assumption that there is a relationship between high counts and slime problems without any scientific confirmation. This means high levels of slimicide might be used to keep counts low, when lower levels

might be sufficient to eliminate slime problems.

A plate count involves taking a water sample from the system, mixing it with nutrients and agar to make a gel, then incubating it for a specified time. Each organism growing well on this medium will produce a colony of visible bacteria which can be counted. The number of colony-forming units per milliliter identified in the sample are then recorded.

On some machines there may be a good correlation between counts and slime problems, not so on others. Before plate counts are used as the only method of monitoring, a good correlation between counts and slime problems should be established. To illustrate this point, Figures 5 and 6 detail plate counts from two paper machines in a coated fine paper mill, along with a count of the number of holes per day in the sheet from each machine. Of course, there are a variety of possible causes of holes in the sheet, not all are slime-related. Holes are a common problem resulting from excess slime growth.

In Figure 5, paper machine #1 reveals an inconsistent correlation between plate counts and holes. The average plate count numbers more than doubled from 11/87 to 12/87, with an insignificant change in number of holes. The average plate count numbers from 2/88 to 4/88 are a small fraction of the numbers from 11/87 to 12/87, and the hole count has dropped somewhat.

In Figure 6, paper machine #2 reflects a significant drop in the average plate count during the second three-month period as compared to the previous three months. However, the hole count changes very little. Two possible reasons account for this. First, it is possible that slime problems causing holes and the plate counts have no relation to the actual slime on the machine. Second, it is possible the holes are related to some

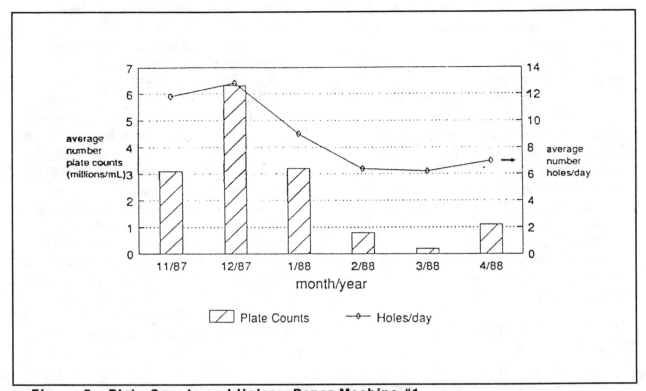

Figure 5.   Plate Counts and Holes - Paper Machine #1
Data from a U.S. fine paper mill indicates how there is often not a close correlation between plate counts and runnability on a paper machine. Also see Figure 6.

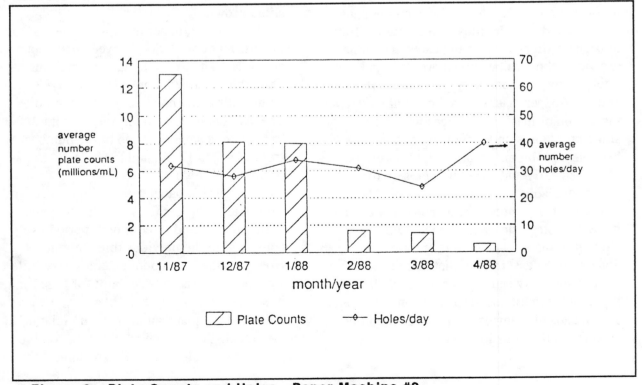

Figure 6.   Plate Counts and Holes - Paper Machine #2
These data are taken from a second paper machine in the same mill as in Figure 5. In this case, even though plate counts change dramatically, slime growth (as determined by hole counts) does not change.

mechanical problem. If this is the case, the plate counts again offer no useful information.

Problems Using Plate Counts to Monitor Slime

1. There may be no correlation between counts and slime problems.

2. It may take more slimicide to control counts than to control slime.

3. Slime-formers may not grow on plates.

4. It is often difficult to reproduce data.

To summarize, do not use plate counts as a monitoring tool unless a correlation between plate counts and slime-related problems has been proven to affect runnability and quality over an extended observation period.

Plate counts can, however, be a tool in a total slime control program. It may be useful to periodically make counts on various parts of the system to make certain there have been no major changes in the number and kinds of microorganisms in various parts of the system. For example, the following parts of the system might be checked periodically: the raw water, the broke, the saveall, coatings, and additives such as starch, dyes, alum, retention aid, etc. Counts may be useful in locating and treating the major sources of contamination. If counts are 20 times higher in the broke than in any other part of the system, it may be wise to treat that area separately with a slimicide.

Devices exist to measure actual slime buildup. One method involves diverting white water through a container with plastic or metal removable plates. At specified time intervals, the plate is removed and the slime scraped off and weighed. If the normal buildup is, for example, 5 grams per unit area of surface, and it suddenly increases to 15,

the operator should be alerted to watch for possible slime problems.

There are a number of other ways to monitor a slimicide program. Some methods use high-tech instruments, others use plastic strips to sample the machine system, others use dyes that change colors as the slime producers grow. All methods have advantages and disadvantages. In every case, before choosing a monitoring method, a correlation between the test and the bottom line should be established. Observe the runnability on the machine and the quality of the sheet, as related to slime problems. The machine surfaces are usually very good indicators of the condition of the machine.

What to Do When Slime Appears

Following are suggested steps for locating sources of slime problems on a paper machine.

Verify Slime as the Problem

Before treating a problem as slime-related, slime should be verified as the cause of the problem. Are the symptoms and location of deposits typical of previous slime problems on the affected machine?

Any deposits should be examined on-site with a microscope by a trained microbiologist. Stains and spot tests should be performed to analyze any troublesome deposits. Occasionally, it is possible to analyze the area around a hole or the breaktail and find evidence of slime or another substance.

Occasionally, deposits appear which look like slime but are not caused by microorganisms. Alum, stickies, or problems with retention can be the cause of slime-like deposits. If a slime problem is not strongly suspected, and the slimicide used has a long history of working well in the mill, a temporary increase in the use rate of the slimicide might offer positive results. If the problems subside, it would indicate the problem is indeed caused by slime-forming microorganisms.

Check Slimicide Feeding Mechanisms

Check to make certain the slimicide pumps are operating correctly. Is the container empty? Are the feed lines in proper working order? Mill or supplier personnel should frequently check levels in the slimicide container making certain the level is dropping and that the product is indeed being pumped into the system.

Temporarily Increase Slimicide

Occasional flare-ups of slime problems in a paper machine prompt operators to temporarily turn up the slimicide feed rate. A typical increase might be 10-20 percent. This often will solve the problem. If the problem recurs when the feed rate is returned to its original position, changes in the slime control program should be considered.

Check Machine Changes

Slime problems occur as a result of many changes on the machine. There may have been a change in flows necessitating a change in the feedpoint for the slimicide. There may have been a change in temperature, pH, or something else at the point where slimicide is fed into the system. A change in additives can also increase the contamination on the machine.

Check Machine Temperature and pH Conditions

The nature and intensity of slime problems vary with climatic conditions. In winter, for example, the white water temperature in some paper mills in temperate areas might drop 15-20 degrees. This may improve the living conditions for bacteria so much that the slimicide can no longer control their growth. It may be necessary to adjust the amount of slimicide used, or increase the temperature of the system by adding steam. In other mills, colder temperatures may encourage fungal growth. If the slimicide used is not effective on fungi, slime problems may appear. It may be necessary to change slimicides in this case.

As discussed earlier, machine pH affects slime problems in two ways. First, bacteria grow much better at near-neutral conditions. Second, many slimicides are not as effective at near-neutral conditions. The pH of the wet end must be monitored as well.

Problems in many mills stem from lack of maintenance of equipment that monitors or controls pH. All pH meters should be cleaned and calibrated according to manufacturer recommendations. Otherwise, the machine pH may be miscalculated.

How Long Since the Last Machine Boil-out?

In most cases, the slimicide program does not eliminate the need for boil-outs. The slimicide may extend the time between boil-outs. Increasing slimicide levels may even extend this time period. At a certain point, however, it is necessary to shut down and conduct a thorough boil-out of the system.

Check Condition of the Fresh Water

A slime outbreak is frequently traced to the fresh water entering the mill. Often the chlorinator has malfunctioned and there has been no chlorine treatment of the fresh water. Although chlorine does not eliminate microorganisms in the fresh water, it does greatly reduce the number of organisms entering the mill. If there is no chlorination of the mill supply water, it may be necessary to increase slimicide levels on the paper machine, or add a slimicide to the mill supply water.

The condition of water from rivers and lakes changes with the season in many mills. Runoff from snow melt or spring rains, seasonal turnovers in lakes, and local agricultural practices may affect water quality and change the number and type of organisms entering the mill system.

Have Fiber Sources Changed?

A number of microorganisms in the machine system, and food for these organisms, come from the fiber raw material used in the machine. Some slime outbreaks may be traceable to this raw material. For example, if high-density pulp has been stored for a longer period than usual, growth of slime-formers in the pulp may result and increase problems on the machine. The amount of slimicide used to preserve pulp in storage might be changed depending on the storage time of that pulp.

Changing from one type of secondary fiber to another, or changing suppliers on wet-lap pulp might change the potential of slime problems on the paper machine.

Have Additives Changed?

The importance of controlling microbiological growth in additives used on the paper machine is clear. It follows that a change in additives used, or in the amount of an additive used, can affect slime growth on the paper machine.

Operators should check the condition of starch, fillers, and broke. Is there any odor or pH drop to indicate slime problems? Starch may need to be treated before or after cooking, sometimes both. Clay may be treated before it gets to the mill, but it often needs additional treatment after a certain

period of time. Occasionally, there are problems in alum, size, or retention aid systems.

Slime Problems in Satellite Systems

Operators should monitor conditions in savealls, and storage chests for white water or stock. If there are indications of slime problems, such as deposits or odor, these systems should be boiled out. If slime problems persist, a slimicide should be used to control the problem.

Contact Slimicide Suppliers

This should be one of the first steps to solving the problem. The manufacturers and suppliers of slimicides have valuable experience in directing proper usage. From the first trial of the slime control program and throughout its implementation, communication between supplier and paper mill personnel is extremely important.

Change the Slimicide Program

If all of the actions suggested thus far do not satisfactorily correct the problem, the basic program may need some adjustment. This might involve a permanent increase in the feed rate, a change in the point of addition, a change in products, or a change in the feeding schedule, for example, from one hour on, three hours off to two hours on, six hours off.

Examples of Slime Control

In this section a variety of situations related to slime control in paper mills will be covered. Some of these situations may provide ideas on improving slime control in a mill experiencing problems.

Problems with Fresh Water

The fresh water used in a mill is a major source of contamination. If the slime deposit appears stringy, problems with the fresh water are suspected since the typical bacteria carried in with fresh water grow in filaments (see Figure 2). The frequency of this type of problem is tied to the frequency of problems with the chlorination system used to treat fresh water. However, even if the chlorinators are working, chlorine is not especially effective on these filamentous bacteria.

The quality of fresh water will vary with its source and with other factors, such as the season of the year. Uncontaminated well water may contain a maximum of a few hundred organisms per milliliter. Water taken from a deep well and then treated with chlorine may contain only 5-10 organisms per milliliter. Surface water contains a much larger population: a pristine stream may have 400-500 organisms per milliliter. A typical river supplying a paper mill may contain a few thousand organisms per milliliter, a lake somewhat less. Spring runoff will affect this number a great deal. Water in a rural stream may have 10,000 organisms per milliliter during the runoff with many of these being slime-forming organisms from the soil.

If a mill uses 5 million gallons per day of water containing 1,000 organisms per milliliter, 800 billion organisms per hour will enter the mill in this way.

Effects of Temperature on Slime

Slime deposits are not a serious problem when paper machine temperatures are around 125°F (52°C) At this temperature, slime control programs typically prevent production or quality problems. However, when temperatures drop, to 100-115°F (38-46°C), it is very difficult to control slime. Levels of slime control products may increase by 25

percent or more. In the example of one paper mill, slimicide usage was increased on several machines as soon as the temperatures began to drop. On other machines the use rate remained unchanged. After two weeks, the machines with an unchanged slimicide use rate showed very heavy buildup of slime in tray areas, while slime remained under control on the other machines. One alternative solution involves using steam, if there is enough available, to maintain higher temperatures, but this is costly as well.

This problem was also observed at another mill. If the wet-end temperature drops, as it occasionally does, from 127° to 115°F (53-46°C), a significant increase in slime deposits results. When this change in temperature is anticipated, mill personnel at one mill routinely double the amount of slimicide used on the machine.

In another mill, tapioca slime appeared suddenly after an effective slime control program had been in place for a long time. After the problem was investigated, it was discovered that changes in the system had contributed to the slime problem. The shower water on the wet end, and some of the seal water, had originally been at 120-130°F (49-54°C). At the time of the slime outbreak, it was discovered that the source of this shower water had been changed to mill make-up water at a temperature of around 90°F (32°C). In addition, for some time the mill make-up water had not been treated with chlorine due to problems with the chlorination system.

Slimicide in a Deinking Plant

The following sequence of events occurred in the deinking plant of another mill when the use of slimicide was eliminated. This example illustrates the effect and the cost of maintaining higher temperatures. The addition point for the slimicide had been into the shower on a gravity strainer used to remove fiber from white water. The filtered water was then reused as shower water. Within two weeks of removing the slimicide, slime began to plug the knockoff showers in the disk saveall. It was also necessary to use additional cold water to keep the strainer clean. This in turn necessitated the use of hot makeup water in the white water system. The cost of the hot water was roughly 13 times the cost of the slimicide needed to control the problem. The mill returned to using the slimicide at the screen.

Fungal Slime Growth at Reduced Temperatures

In other areas, temperature changes have different effects on slime growth. During winter, mills located in some temperate areas experience dramatic temperature drops on the paper machine. A change in the type of microorganism growing on the machine is commonly observed. As the temperature drops to approximately 120°F (49°C), fungal growth appears. Fungi cannot compete well with bacteria at the higher temperatures, but can be a serious problem when cooler temperatures prevail. It may be necessary to change the slimicide used at this time of year, since many slimicides are not effective against fungi.

Problems Related to Anaerobic Bacteria

The problem of odor-causing and poison gas-producing bacteria is more prevalent in conditions where no oxygen is present. These bacteria, called anaerobic organisms, can cause "sour stock," the buildup of hydrogen or other explosive gases, and can cause corrosion of stainless steel and other metals.

Following are suggestions on controlling problems caused by anaerobic bacteria. Several of these are designed to maintain

quick turnover of stock in the system, and so to prevent areas where oxygen levels are depleted.

1. Maintain good agitation in chests where stock or white water is stored; avoid dead-flow areas.

2. Keep water and fiber reuse circuits as short as possible.

3. Keep retention times in the various stages to a minimum.

4. Keep volumes of stock and water in various chests to a minimum (consistent with adequate surge capacity).

5. Use flotation savealls or clarifiers using air flotation to help maintain oxygen levels.

6. Maintain effective slime control and control of other deposits.

7. Maintain a clean paper machine system with regular maintenance and effective boil-outs.

Slime in Storage Chests

Chests used for pulp and water storage are often neglected in a mill. In some cases, handfuls of slime can be removed from these chests. Often, when the chests are drained, an area of gray or black stock remains at the bottom of the chest. This is another indication of anaerobic bacteria growth which can produce dangerous gases, degrade fibers, cause odors, and cause corrosion.

Recently there has been more interest in controlling slime in these areas, since some mills have had explosions caused by bacteria growing in chests. Potential solutions include the following:

1. Regularly scheduled washups in chests may not be feasible, but should be considered.

2. Always maintain good agitation in the chests.

3. Apply slimicide to control growth of microorganisms.

4. See recommendations in the previous section,"Problems Related to Anaerobic Bacteria."

Change in Structure of Saveall

Following is an example of what a mill might do to reduce buildup of anaerobic bacteria in the system. In one case, the problem was traced to the flotation saveall system. The major cause of the problem in this case was the design of the saveall. The overflow tank , a round vessel with a flat bottom and no agitator, is used for level control in the saveall system. The stock is pumped out on one side of the tank near the bottom. This creates an area which is seldom mixed or turned over thus becoming a perfect area for anaerobic bacteria to grow and cause odors, discoloration, and corrosion.

In this mill, after two weeks of running, there would be four feet of rotten stock in the bottom of the tank on the side away from the pump. After 4-5 days of bacteria growth, bubbles of gas could be observed coming to the surface. It was necessary to hose out this tank frequently. The buildup in the bottom of this tank was a major source of fouling to the rest of the system.

In this case, the best solution was to change the design of the tank. A sloping floor was added to the tank thus eliminating the area that could not be agitated. The problem of bacteria growing here was eliminated. This change provided the added benefit of removing a source of contamination to the machines, since the solids collected in the saveall are returned to the wet end of the paper machines.

Effect of Nutrients on Slime Growth

A change in the level of nutrients may be the cause of a sudden flare-up of slime problems. This might be due to a seasonal occurrence such as spring runoff. Aside from that, a change in additives can alter the level of nutrients in the system. A change in type of starch used might increase the level of nutrients. A change in type of sizing, for example, to a material containing nitrogen can cause a sudden increase in slime. A change in the chemical treatment of the incoming fresh water may increase levels of phosphate, another essential nutrient. A change from caustic to ammonium hydroxide to control pH in starch can cause a slime problem in that part of the system.

Effect of pH on Slimicide

Following is an example of a case where a small adjustment in the pH of part of the system affected slime control. In a large kraft linerboard mill, part of the slime control program involved the addition of a slimicide to the furnish at the machine chest. The major problem in the mill, when effective slime control was not in use, was the buildup of tapioca slime in the cleaners and headbox. For a long time slime had been well under control, but problems suddenly appeared. After some careful study of the system, it was determined that the pH at the machine chest had risen substantially, from around pH 5 to pH 9. The pH on the wet end of the paper machine had not changed at all, since pH is controlled after the slimicide addition point. However, due to variations in washing in the pulp mill, the pH at the machine chest had risen.

The problem here was not the slimicide, but the change in the system. Most slimicides currently in use will degrade quickly after addition to the system. However, many products degrade extremely quickly at high pH levels. In this particular case, the slimicide was destroyed very soon after addition to the furnish, and the bacteria were not controlled. To eliminate the problem, operators could either change the pH at the machine chest, or change the slimicide addition point.

Machine Deposits Other Than Slime

The following problem occurred in a fine paper mill that had converted to producing a neutral-sized, calcium carbonate-filled sheet. The mill was experiencing runnability problems and frequent breaks. The wet end was very sloppy and the problem was thought to be slime. In fact, the theory was that the slime was due to contaminated lap pulp. However, microscopic examination of the deposits revealed no bacteria or fungi present. The samples consisted of about 95 percent calcium carbonate. The problem was a buildup of filler on the wet end, possibly due to a problem with retention, starch, or sizing, but not due to slime.

Chlorine and Related Chemicals

The use of chlorine, hypochlorite, or chlorine dioxide is often an important part of a slime control program. These materials have been available to the industry for 50 years or more and are effective on some types of bacteria. These oxidizing slimicides are quite effective, for example, on pathogenic (disease-causing) microorganisms. However, the products are not especially effective in controlling fungi or filamentous bacteria.

In some mills, the use of these products on the wet end of the paper machine works quite well. In other mills, chlorine and related chemicals have little or no effect. In many mills, chlorine treatment of the mill supply

water is an important part of keeping slime under control.

One tissue mill runs with a wet-end pH of about 6.5-7.5. When chlorine dioxide alone is used for slime control, there continues to be slime problems from the growth of fungi and filamentous bacteria (see Figure 2). Adding a conventional slimicide to the machine eliminates the slime problem.

In another mill, conventional slimicides had been used for many years to control slime in machines used as dryers for highly bleached draft pulp. In addition, chlorine dioxide was added to the white water storage chests. Chlorine carryover from the bleach plant also helped control slime on these machines. For many years this program worked well, but slime problems suddenly appeared. In this case, plate count was another indicator of a problem. The counts began to average five times higher than normal.

Investigation of the system turned up the fact that the mill had begun adding a chemical into the system to kill chlorine in order to eliminate odor. In this case, the use of the conventional slimicides had to be doubled to counteract the elimination of chlorine from the system.

Experiences with Bacteria-Related Explosions

Explosions related to bacterial activity have occurred in some paper mills. In one Canadian mill, investigators found significant amounts of hydrogen were produced in white water storage chests within 10-12 hours during a machine shutdown. Hydrogen accumulates when stagnant fibers trap gas, or when there is little ventilation to disperse gas. Here again, it is important to provide agitation.

Apparently these hydrogen-producing bacteria grow much better in alkaline conditions, and do not thrive when the pH is lower than around 5.5. As more mills run at alkaline pH and reuse more water (close up the system), conditions are more favorable for this type of bacteria.

Costs and Benefits of Effective Slime Control

When the growth of microorganisms is severe enough to cause breaks on the machine or require an unscheduled shutdown for cleaning, preventative measures must be taken. The most expensive problem on a paper machine is loss in production. In some mills down-time is calculated at around $5,000 per hour; in others, up to $25,000 per hour, including the cost of the paper not produced.

Problems of lost production time can be broken down as follows:

1. Cost of segregating and repulping rejects.

2. Cost of rejects not repulpable.

3. Loss from customers claims.

4. Loss in weight and paper quality due to overdrying which is done to get uniformity and meet customer specifications.

5. Customer rejects due to defects in product.

6. Loss of customers to competitors.

Consistent high quality is becoming even more important in today's market. The mill turning out a sheet marred by spots, holes, or odor due to microorganisms will be quickly out-distanced by competitors.

In summary, slime is a very costly problem to the paper industry. However, with common-sense strategies and proper use of available chemicals, any slime problem can be controlled.

Sources of Additional Information

1. Purvis, M. R. and Tomlin, J. L., "Microbiological Growth and Control in the Papermaking Process," *TAPPI Chemical Processing Aids Short Course*, (Seattle) April 10-12, 1991, pp. 69-77.

2. Hoekstra, P. M., "Fundamentals of Slime Control," *TAPPI Chemical Processing Aids Short Course*, (Seattle) April 10-12, 1991, pp. 55-68.

3. King, V. M., "Microbial Problems in Neutral/Alkaline Paper-Machine Systems," *TAPPI Neutral/Alkaline Papermaking Short Course* (Orlando) October 16-18, 1990, pp. 211-216.

4. May, O. W., "Chemical Processing Aids in Neutral/Alkaline Papermaking: Overview," *TAPPI Neutral/Alkaline Papermaking Short Course* (Orlando) October 16-18, 1990, pp. 197-199.

Chapter 2

Agents for the Control of Pitch, Scale, and Other Nonmicrobiological Deposits

by Alan J. Schellhamer, Wallace E. Belgard,
and George S. Thomas

Introduction

Most pulp and paper mills experience some form of chemical deposition on process equipment which, if not controlled, can result in a variety of problems including reduced production rates, increased maintenance and cleaning costs, increased energy costs, and product quality defects. Some chemical deposit problems are a result of equipment and process limitations or improper process and chemical control procedures which, if optimized, can reduce or eliminate the need for deposit control additives. However, many chemical deposit problems are a natural by-product of the basic raw materials and chemical operating conditions which must be utilized in the manufacture of pulp and paper. The focus of this chapter is on the nature and application of those products generally described as deposit control agents which have been developed by various suppliers to control most types of chemical deposit problems. Products used to enhance the cleaning effectiveness of system boil-outs are also included.

Description, Function, and Application of Deposit Control Agents

There exists an extremely wide variation in the chemical composition and function of deposit control agents. However, it is possible to classify and describe these products on the basis of the chemical mechanisms by which they function to control specific classes of deposits such as pitch, scale, general chemical debris, and stickies.

Pitch Stabilizers
Description and Function

Pitch stabilizers are typically liquid formulations of nonionic and anionic surfactants. These products are water-based, having low viscosity and low corrosivity and ranging in pH from mildly acidic to alkaline. As a result, they are easily transported, stored, and metered using materials of handling recommended by the supplier. Some pitch stabilizers are supplied in powdered form and therefore must be made down in water prior to use.

Both types of surfactant molecules in pitch stabilizers contain a lipophilic or fat-loving portion (also described as organophilic or hydrophobic) which adsorbs onto the similarly lipophilic portion of pitch molecules or the surface of pitch particles. The hydrophilic (water-loving) or water-soluble portion of the surfactant molecule can be either nonionic (uncharged) or anionic (negatively charged). The hydrophilic portion of the molecule remains oriented into the surrounding water phase. In effect, this results in an electronic or physical barrier, thus reducing the tendency for pitch molecules and small particles to come together to form larger aggregates, to form a film at an air-water interface, or to adhere to an equipment surface. The terms dispersion and wetting are most commonly used to describe this action of pitch stabilizers.

Surfactants, including those used in pitch stabilizer formulations, have a tendency to stabilize entrained air bubbles in water and thus to aggravate foaming conditions. However, foaming is normally not a problem in pitch control applications due to the relatively low application concentrations used.

Product Selection and Application

As with most deposit control agents, pitch stabilizers are normally developed to control specific chemical species of pitch under the particular chemical conditions found in the pulping and papermaking process in which they must function. Therefore, the primary basis for the selection of the most appropriate pitch control agent in

any given situation is the chemical composition of pitch deposition to be controlled and the chemical conditions of the system in which the product is to be applied. In order to properly design a pitch control program, samples of pitch deposits throughout the system must be obtained for thorough chemical analysis, and the system itself must be analyzed to determine all chemical and operational factors which will have an effect on the performance of the pitch stabilizer applied.

Because the chemical composition, physical form, and location of pitch deposition varies widely among the various pulping, bleaching, and papermaking processes, there are a variety of pitch stabilizer application points used to take advantage of one or more of the control mechanisms of a given product. These application points will be selected by the supplier based on their analysis of the system and experience in other mills.

Product feed rates are normally specified in terms of pounds of product per ton of pulp or paper production and converted to milliliters per minute for purposes of feed pump metering. The supplier's recommended initial feed rate is based on a number of factors in each specific application and also on the feed rate found to be effective in similar applications in other mills.

The ultimate effectiveness of a pitch stabilizer application should be evaluated on the basis of its ability to significantly reduce those operating or quality problems which created the need for treatment. This is accomplished by identifying the pretreatment average values of appropriate production, quality, or cost statistics against which treated values are compared to determine significant changes. Because of the great number of variables unrelated to pitch control which often affect production and quality, a treatment evaluation period of up to three months is often desirable. However, intermediate evaluations of treatment effectiveness are made by periodic inspection of selected equipment areas or by measuring deposition on pitch plates placed in historically high-deposition areas of the stock stream.

Pitch Detackifiers
Description and Function

Pitch detackifiers, a recent introduction to the pulp and paper industry, have gained acceptance very quickly. These pitch detackifiers are liquid formulations of nonionic organic polymers which, as with nonionic surfactants, contain both lipophilic (hydrophobic) and hydrophilic functional groups. Pitch detackifier polymers function much like nonionic surfactants to make pitch particles less hydrophobic and prone to deposit or agglomerate. However, due to their polymeric nature, pitch detackifiers have the ability to coat pitch particles with a multi-molecular layer of polymer. This multi-molecular polymer coating acts to insulate the tacky hydrophobic surfaces of pitch particles to prevent their agglomeration or deposition. They are typically easy to handle, moderately viscous, non-corrosive solutions which are easily transported, stored, and metered using materials of handling recommended by the supplier.

Pitch detackifier polymers are well-suited for pitch control in stock prep and machine systems. They are nonionic, so wet-end chemistry balance is not affected. They also

tend to be completely exhausted onto pitch particles, leaving little or no residual in the water at typical dosages. They also have low foaming tendencies and can control pitch over a wide pH range. Pitch detackifier polymers have also found use in alkaline pH machine systems where control of hydrolyzed sizing agents is a concern.

Pitch detackifier polymers are also useful in controlling pitch deposition in the pulp mill and bleach plant.

Product Application Technology

Detackifier product feed rates are normally specified in terms of pounds of product per ton of pulp or paper production and converted to milliliters per minute for purposes of feed pump metering. The suppliers' recommended initial feed rate is based on a number of factors in each specific application and also on the feed rate found to be effective in similar applications in other mills.

The effectiveness of pitch detackifier polymers should be evaluated just as that of other pitch control agents. Visual inspections of process equipment, weighing of pitch plate scrapings, and tabulation of the economic cost of a pitch problem in terms of machine runnability and finished paper quality are typical measurements of the success of a pitch control application.

Pitch Adsorbants
Description and Function

Pitch adsorption with talc represents an alternate and often effective method of pitch control. Talc is a powdered form of magnesium silicate which exists in the form of delaminated, plate-like particles with an average size of one to two microns and a high-pitch adsorptive surface area. Talc particles function by adsorbing pitch molecules, filmy pitch, and colloidal pitch particles on their organophilic (lipophilic) surfaces. Following adsorption, pitch is carried through the stock system on the talc particles which can be incorporated into the finished product or they can leave the system with the effluent. End product retention of talc varies widely depending upon feedpoints.

Maximum effective utilization of talc depends upon feeding at one or more points in the process where pitch exists in either molecular, filmy, or colloidal particle form. Talc particles are capable of coating larger pitch particles but, in this form, are sensitive to shear. This can result in downstream exposure of the tacky pitch particle surface and deposition. As a result, talc is normally fed at one or more points in the process.

Talc Handling and Feeding

Talc is supplied in a low-density, free-flowing powder form or a high-density,

compacted powder or pellet form and normally packaged in 50-pound paper sacks or 1,000-pound plastic bulk sacks. Although the low-density powder form requires less vigorous agitation and, therefore, less equipment to slurry, the compacted form costs less to ship, requires less storage space, and produces less dusting when the shipping sacks are emptied.

Talc is normally fed into the process system as a slurry containing one-half to one pound of solids per gallon of water and at a point of agitation for optimum distribution in the stock. The slurry is most often prepared in a talc dispersing unit (see Figure 1). The unit consists of a hopper which can be designed to hold 1,000 pounds of compacted talc, a variable-speed screw feeder, an initial dilution cone with water supply, and three successive agitation tanks. The consistency in the first two tanks is normally maintained at two pounds per gallon. The third tank acts as a surge tank and final dilution point to adjust the slurry to any desired final solids content. Depending upon the location of the dispersing unit, the talc suspension may be

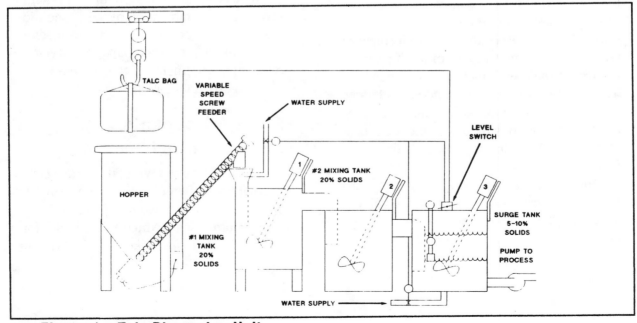

Figure 1. Talc Dispersing Unit

fed to the process by gravity, or by means of a positive displacement or centrifugal pump.

For convenience, powdered talc is sometimes fed intermittently to the stock system in dry form. Although adequate for batch pulper addition, this method of feed is not recommended to stock storage chests as it results in poor talc distribution and inefficient usage.

Pitch and Stickie Passivators

Description and Function

A new type of specialty chemical application for controlling deposition of pitch, stickies, and other hydrophobic materials on paper machine fabrics entails the use of a water-soluble, low molecular weight cationic polymer. This polymer is diluted in shower water and sprayed directly onto the forming fabric. It is thought to function by adsorbing a thin cationic layer onto the surface of the fabric. This cationic layer then attracts various anionic trash to its surface so that a thin layer of material is formed on the surface of the fabric which then passivates or detackifies the fabric to prevent the adherence of pitch, stickies, or other hydrophobic materials. This thin layer of polymer and anionic material is thought to be responsible for the characteristic tan to brown color observed on fabrics using this treatment program. These type of programs can result in efficient, cost-effective control of fabric filling.

Product Application Technology

Passivation products are typically fed through a fan type shower onto the surface of the fabric. The product supplier's recommended initial feed rate is based on a number of factors in each specific application and also on feed rates found to be effective in similar applications in other mills. These products can be evaluated by comparing the rate of fabric filling in pre-treatment phase with the rate of filling after treatment has begun. Other runnability and quality parameters such as sheet spots, holes, and runnability can be monitored in order to determine the program's efficacy and return on investment.

Dispersants for General Chemical and Particulate Debris

Description and Function

Excessive quantities of sludge-like deposit masses formed on the paper machine wet end often lead to production and quality problems. These deposits may consist of varying amounts of acid or alkaline size components, pitch, fillers, cellulose, and alum. The deposits normally contain some percentage of biological matter. However, when the biological content is relatively low, these deposits are generally not considered to be primarily of biological origin. In these cases, effective stock washing, careful control of paper machine chemistry and additives, and effective routine boil-out and wash-up procedures can be practiced to minimize deposition. However, when deposition is rapid, dispersants can aid machine cleanliness, runnability, and product quality.

General wet-end dispersant products are normally liquid formulations consisting of single components or combinations of nonionic and anionic surfactants or polymeric, anionic dispersants. These products are water-based, low-viscosity, low-corrosivity formulations ranging in pH from slightly acidic to mildly alkaline. As a result, they are easily transported, stored, and metered using materials of handling recommended by the supplier.

Surfactant components of general dispersants are designed to wet and disperse lipophilic furnish components such as pitch, oil, and acid or alkaline size products which often act as binders for other materials in the deposit. These components also act to penetrate and disperse the gelatinous biological components of deposits. Anionic dispersant components in these products are designed to adsorb onto the surface of particulate matter such as filler, pigments, and precipitated salts. The dispersant acts to increase the negative surface charge of particulates, and decrease their tendency to form large aggregates or to settle out onto equipment surfaces in low-flow areas.

Product Selection and Application

The primary basis for selection of a general wet-end deposit dispersant is the chemical composition of representative deposit samples.

Effective product feed rates and feedpoints vary substantially from system to system due to the wide variety of papermaking chemistries and furnish components. The initial trial feed rate and feedpoint is usually based on a supplier's treatment experience in other systems producing similar paper grades and is adjusted for the severity of deposition in a particular system.

The most common method of general wet-end dispersant evaluation involves visual inspection of key wet-end areas for deposit coverage, mass, and tenacity. Deposit samples should also be obtained during the evaluation period to determine changes in the relative amounts of various deposit components. On-site microscopic and chemical analysis capabilities are normally available from suppliers to help make adjustments in product feed rates to obtain the most effective and lowest cost total wet-end deposit control program.

Production and quality records can be utilized to critically evaluate the impact of the application.

Scale Inhibitors
Description and Function

Most scale inhibitors are liquid formulations of one or more types of anionic polyelectrolytes which carry functional groups including, but not limited to, phosphate, phosphonate, sulfonate, or carboxylate. These products are water-based solutions, having relatively low corrosivity and ranging in pH from mildly acidic to mildly alkaline. As a result, they are easily transported, stored, and metered using materials of handling recommended by the supplier. Some scale inhibitors are supplied in granular form and, therefore, must be dissolved in water for metering to the injection point.

The principle mechanism used to control such common scales as calcium carbonate, calcium oxalate, and barium sulfate is called inhibition. In this case, much lower than stoichiometric quantities of scale inhibitor molecules function by adsorbing onto the surface of initial scale-forming clusters of ions which start to form on an equipment surface. The high anionic charge imparted by the inhibitor repulses the anionic depositing species (carbonate, sulfate, or oxalate), thereby interrupting the scaling process and reducing the rate of scale growth. After crystallization has begun, these inhibitors can adsorb onto and modify the surface of the crystal to prevent crystal growth and subsequent scale formation.

Product Selection and Application

Proprietary scale inhibitor formulations are designed to control specific scaling species under chemical conditions which best simulate the process environment in which scaling occurs.

Each type of scale necessarily forms under a specific set of chemical conditions which exist in certain pulping, recovery, bleaching, and papermaking systems. Therefore, products and feedpoints are selected based on analysis of the system and supplier experience in other mills.

The required treatment dosage for a given type of scale can vary substantially from system to system and at different times within the same system. Factors such as depositing anion concentration, pH, temperature, total cationic species concentration, conductivity, and total suspended solids surface area all have an effect on the required dosage. Based on laboratory test data, treatment experience under a variety of conditions in other systems, and an analysis of the chemical factors in the system to be treated, the supplier recommends an initial product feed rate. However, it is normally necessary to increase or decrease the feed rate based on the degree of control achieved during the initial evaluation period in order to optimize control

or cost. Feed rates are normally specified in terms of pounds of product per ton of pulp or paper production but may also be specified by ppm concentration for application to liquid streams.

As with other deposit control agent evaluations, the effectiveness of a scale control program is ultimately evaluated by the degree to which it reduces or eliminates the production or quality problems which have been identified as being caused by the scale. Statistical comparison of appropriate pretreatment production, quality, or cost data to treated system data is the most definitive means of evaluation. Depending upon the nature of the problem, evaluation periods of one to six months or more may be required. Also, because scale control agents are unable to directly remove scale over that time span, it is important to remove as much scale as possible prior to the evaluation by chemical or mechanical cleaning in order to make accurate judgments about scale inhibitor effectiveness during the evaluation period.

Intermediate treatment evaluations can be made by periodic inspections of scale-prone equipment surfaces. Photographs of key areas prior to and during treatment are helpful to maintain an accurate, historical record of scaling conditions.

Boil-out Additives
Description and Function

Boil-out additives are cleaning compounds used to remove system contaminants and deposits. Typically, these compounds are used in the paper machine back systems, as well as wet-end and starch systems. Boil-outs are usually done during a scheduled shutdown as a part of the total maintenance program. The frequency of boil-outs is specific to each machine and the grades produced.

Boil-out additives are supplied in either liquid or dry (granular) forms. These products may contain several cleaning compounds, but are generally classified as either caustic, neutral, or acid products. Caustic or acid-based products are normally intended to be used as a single source of all necessary boil-out chemical requirements. Most frequently, the neutral products are intended to be used in conjunction with open market caustic or acid.

Cleaning compounds normally included in boil-out products fall into three categories of chemicals as follows:

1. Sequestering or chelating agents are used to increase solubilization of scales such as barium sulfate. They are also used to sequester cations such as calcium, magnesium, and iron.

2. Detergents and emulsifying agents function to penetrate, loosen, and disperse organic deposit binders such as pitch, size, oil and other hydrocarbon-based materials.

3. Solvents may be used to soften or dissolve particularly tenacious organics such as latex, resins, and other stickies.

Powdered, granular, and liquid products are normally added directly from the shipping containers to the recirculating boil-out water. When handling the boil-out products, maximum safety precautions should be employed. Safety goggles or a face shield, rubber gloves, and protective clothing should be worn at all times. Special care should be taken to avoid inhaling dust while handling powdered products.

Product Selection and Application

Because of the specific nature of each class of cleaning compound, selection of an appropriate boil-out program is based on an analysis of the deposits to be removed. Simple laboratory deposit solubilization tests may also be performed to screen candidate boil-out additives and to estimate the required boil-out concentration.

The required concentration of boil-out additives is normally specified in terms of pounds or gallons per 1,000 gallons of boil-out solution and is based on laboratory screening tests or prior experience.

Evaluation of boil-out effectiveness should involve a thorough inspection of all areas exposed to the boil-out solution including normally inaccessible areas such as filtrate tanks, stock chests, and the headbox manifold. Remaining deposits should be obtained and analyzed to compare against the analysis of pre-boil-out deposits. The post-boil-out inspection is also essential to find loosely adhering debris which should be removed with high-pressure water and sewered before machine start-up to avoid breaks and quality defects after start-up.

Deposit Control Agent Problems and Solutions

Using deposit control agents as prescribed by the supplier normally results in minimal feeding or process difficulties. However, there are potential problems

---

Table 1. Problems and Solutions Associated with Deposit Control Agents

Problem	Possible Solution
Loss of product flow	Check: • Feed pump electrical supply • Pump pressure rating vs. discharge pressure • Loss of pump suction • Plugged injection tap, check valves, or feed line • Feed line break or connector leak • Pump motor failure • Pump check valve or diaphragm failure • Product storage container contamination; sludge
Feed line/pump liquid-side deterioration (softening, cracking, pitting, swelling)	• Change to recommended alternate materials of construction
Excessive process foaming	• Check for possible system pH control upset • Reduce target feed rate by 25% • Move feedpoint downstream of excessive agitation • Eliminate source of system air entrainment • Change to alternate product
Suspected interference with or by another system additive	• Increase distance of separation from additives • Ensure good agitation at feedpoint • Check for excessive deposit control agent feed rate • Stop deposit control agent feed to confirm interference

---

associated with deposit control agent use and, in all cases, the supplier should be consulted to determine the cause and corrective action. Table 1 summarizes the most common potential problems and solutions associated with using deposit control agents.

Liquid Deposit Control Agent Storage and Feeding

Storage Containers

Liquid deposit control agents are commonly supplied in drums, intermediate quantity bins of various constructions, and bulk tank truck quantities. The supplier utilizes drum types and bin materials which are compatible with the composition of each product and conform to appropriate government regulations. Bulk storage tanks should also be constructed of liquid-side materials recommended by the product supplier. However, 304 stainless steel with or without a phenolic resin lining, polyethylene, or fiberglass-reinforced polyester tanks are most commonly employed.

The maximum storage time and storage temperature range are specified for each product to ensure problem-free handling and use. Product left in storage longer than the

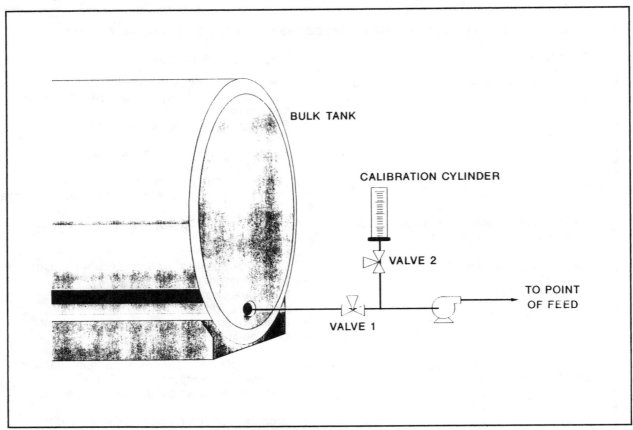

Figure 2.   Bulk System Feed Rate Measurement

recommended storage period should be sampled and given to the supplier for quality-assurance testing.

With the use of drums or bins, a daily or once-per-shift container drawdown measurement is the most simple and accurate method of checking product feed rate. However, the inability to accurately measure bulk tank drawdown on a short-term basis creates the need for an alternate method. One method involves direct measurement of the feed flow, diverted through an accessible line, with a graduated cylinder and stopwatch. However, this method can result in errors due to a possible difference in flow rate between discharge to atmospheric

pressure vs. connected discharge against back pressure. An alternate method, which is generally more accurate and simpler to perform, involves installation of a 100-250 ml graduated cylinder on the suction side of the feed pump (see Figure 2). This method allows fast and accurate measurement of the cylinder drawn down rate while pumping against the actual feed pressure. The cylinder and appropriate valving are mounted so that the top of the cylinder is in line with the lowest normal tank inventory level.

Also available on the market are automatic feed systems which automatically adjust the feed rate of the product based on appropriate process parameters such as production rate,

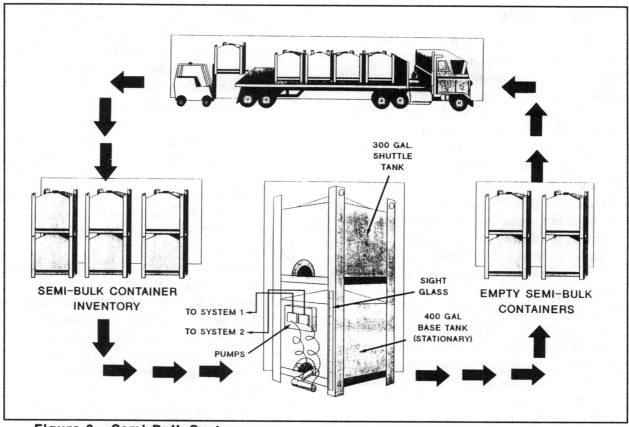

Figure 3. Semi-Bulk System

flow rate, pH, etc. Some of these units can also measure drawdown for inventory control.

Bins which contain volumes greater than drums have more recently come into widespread use. Bins feature the advantage of significantly reducing container handling and change-overs as compared to drums, thus allowing location of the feed containers close to multiple application points, and eliminating bulk tank and lengthy feed line installation costs. One example of this type of system is the semi-bulk system (see Figure 3). This system uses a 400-gallon base

tank equipped with a multiple feed line connection manifold, a sight glass mounted on the manifold for tank level indication and feed rate measurement, and multiple feed pump housing with electrical outlet. The 300-gallon shuttle tanks are designed to be stacked on top of the base unit for rapid product transfer via a ball valve and quick-connect hose. Empty shuttle containers are returned to the receiving dock storage area for pickup and return to the supplier. This eliminates drum disposal problems for the user.

Feed Pumps

Plunger and diaphragm-type positive displacement pumps are the most widely-used deposit control agent feed pumps. This class of pump delivers an accurate incremental volume of liquid. The delivery rate is governed only by the pump speed and is essentially unaffected by suction or discharge pressures. If the discharge line is mistakenly closed, the pump will stall or develop dangerous line pressures. To prevent this, the pump must be protected by a relief valve in the discharge piping or be equipped with an internal relief mechanism.

One pump or pumping head must be used for each point of application because it is impossible to throttle discharge from one pump so that several points will receive a controlled amount of treatment. Satisfactory division of flow is not possible because the discharge pulsates and is normally at a low rate. The use of specially designed and calibrated flow control valves for throttling is also not helpful. Pump heads are commonly constructed of PVC, but may be a variety of other materials compatible with nearly any chemical solution.

The plunger-type pump is commonly used to pump effectively against high discharge pressures. This type of pump operates at a constant rate but the stroke length can be adjusted to vary the delivery capacity of each stroke. With most models, the adjustments must be made while the pump is stopped and the feed rate can be varied from 10 to 100 percent of rated pump capacity.

The diaphragm pump uses a flexing diaphragm rather than a plunger to develop the pumping stroke. The diaphragm is normally Elastomer or Teflon. One pump design drives the diaphragm mechanically with an oil-immersed motor. This pump is constant-rate, but the stroke length or delivery capacity is adjustable from 10 to 100 percent of rated pump capacity and can be adjusted while the pump is running.

Another design uses a solid-state electronic pulsing circuit which drives a linear power coil to provide the diaphragm stroke. This allows adjustment of both stroke length and frequency. The control range of this combination is from 1 to 100 percent of capacity and is adjustable while running.

A third design of diaphragm pump uses an internal hydraulic system to operate the diaphragm. This type of pump is available in models operating at discharge pressure exceeding 1,500 psig. The delivery rate of the pump is manually adjustable while running or can be automatically adjusted by either pneumatic or electric control signal. The latter two diaphragm pump types have the advantage of a built-in pressure relief mechanism.

Regulatory Information

Deposit control agents generally are not a highly regulated class of products. This is due to the fact that the components of these products are not usually hazardous.

The United States Food and Drug Administration (FDA) regulates chemical applications to pulp, paper, and paperboard where the manufactured product is meant for direct contact with food. The FDA approves specific ingredients which can be used in the manufacturing of such paper products. Part 176 of Title 21 of the Code of Federal Regulations (CFR) lists the approved ingredients which can be used as indirect food additives in paper and paperboard. In addition, substances generally recognized as safe (GRAS) by the FDA and listed in Part 182 of Title 21 are authorized to be used in the manufacture of paper meant for food contact. In instances where paper or paperboard products are not manufactured specifically for use with food, the individual manufacturer sets company policy regarding the FDA status of additives used.

The United States Department of Agriculture (USDA) governs application in meat, rabbit, egg, and poultry plants. The USDA will grant approval for products used in the manufacture of paper meant for the packaging of these products. Generally, if the components of a product are FDA approved, USDA will also grant its approval. The USDA lists the approved products in a publication issued every six months by its Scientific Services (APHIS).

The Occupational Safety and Health Administration (OSHA) administers various state right-to-know laws which require material safety data sheets (MSDS) to be issued for each product. The MSDS indicates safety considerations and handling precautions to be used for the product and may always be requested from the supplier.

Sources of Additional Information

There are no other general references presently available regarding the application of all deposit control agents. However, product and application information is available for specific products from all manufacturers and suppliers of these products.

Chapter 3

Papermachine Clothing: Cleaning and Conditioning

by Linda Brooks Bunker
and
Dr. William E. Smith

Introduction

Forming fabrics and wet felts serve to convey the wet sheet from the headbox to the first dryer. During the process, the sheet consistency increases from 1 percent or less to 40-45 percent. It is extremely important that the sheet delivered to the dryer section be uniform in moisture-free weight and moisture content, both in the cross direction and machine direction. Uniformity is the key to sheet quality. Unless the forming fabrics and wet felts are uniformly permeable to the water which passes from the consolidating sheet into them, the goal of a quality sheet will be a practical impossibility.

Not only must the clothing on the wet end be uniformly permeable, but it must also be free of surface contamination in the form of insoluble solid residue that can lead to localized crushing, pinholes, and picking of surface fibers from the sheet. The buildup of finely dispersed solids, such as clay in the interstices of fabrics and in the base yarn regions of wet felts cannot be allowed to

occur. Even though short-term uniformity might be acceptable over a period of hours, or in some cases days, a forming fabric filled in this manner will not drain properly, and the ability of a wet felt to accept sheet water will be impaired. Sooner or later runnability and quality problems will develop.

The yarn properties and designs of the clothing involved must meet highest quality standards to achieve the desired effects. Wet-end clothing must be either continuously cleaned, conditioned or both in most cases. Reasons why emphasis must be placed on clothing cleaning today include the following:

1. Fiber length is decreasing as 100 percent TMP newsprint is produced, hardwood utilization is emphasized, and secondary fiber is introduced to all grades of paperboard.

2. Papermachine speeds are higher, twin wire formers and fourdriniers with top wires are gaining in popularity, and high intensity pressing is

emerging. All of these technologies are more demanding of the clothing.

3. Quality is being emphasized in the global paper and board marketplace.

4. To remain competitive, mills must optimize paper performance. Since the press section removes water more economically than the dryer section, wet felts must be conditioned to remove sheet water efficiently.

Just as uniformity is important in the forming and press section, uniform permeability is a requirement for optimum dryer felt performance. A uniformly permeable dryer felt allows drying to proceed efficiently as water vapor transmission is maximized and the pocket ventilation system operates effectively. Since dryer felts often run up to 400 days, this clothing is subjected to a variety of filling materials from normal operations as well as system upsets. Contamination can lead to picking and loss of heat transfer as well as non-uniformity in drying. It is therefore important to clean dryer felts whenever there is a short outage. If properly planned for and executed, dryer felt cleaning can remove fiber, fines, coating, size, insoluble glue solids, wax, resins, and wood pitch. In many cases up to 98 percent of the original dryer felt permeability is restored.

General Description
Purpose

The primary purpose for using a machine clothing cleaner-conditioner is to enhance water removal in a uniform manner while extending clothing life. Some cleaner-conditioner formulations enhance water removal by increasing the rate of evaporation of residual water in the sheet. As the sheet and wet felt leave a nip there is a small amount of rewetting from the wet felt to the

sheet. Consequently, a small amount of cleaner is transferred to the sheet surface. The cleaner disturbs the hydrogen bonding of water and lowers its surface tension, thereby enhancing the rate of evaporation of water as the sheet is heated in the dryer.

Most wire and felt cleaners function by eliminating solid or viscous materials accumulated in the structural voids, thereby improving dewatering at the foils and uhle boxes, as well as improving the hydraulic flushing action. A given cleaner is designed to attack specific contaminants. Solid or viscous contaminants are usually an accumulation of one or more of the following:

1. Stickies from secondary fiber
   Glue residue
   > Wax
   > Foam
   > Rubber
   > Asphalt

2. Natural wood pitch, colloidal or suspended, from groundwood, semichemical, or sulfite pulp

3. Pitch-like additives
   Coating materials
   > Aluminum silicate (clay)
   > Titanium dioxide (TiO_2)
   > Calcium carbonate ($CaCO_3$)
   > Latex

4. Stock and water additives
   > Defoamers
   > Pigments
   > Talc
   > Size
   > Retention/drainage aids

5. Inorganic precipitates
   > Calcium oxalate
   > Barium sulfate

6. Wood fiber, fines, bark, system dirt

7. Biological deposits

Strong acids or alkalis will often clean the contaminants out of the machine clothing. Specially designed cleaners-conditioners not only free the clothing of solid or viscous contaminants but they help flush out abrasive fillers thereby extending clothing life.

Description

Clothing cleaners remove or prevent deposits by three mechanisms: emulsification, water-phase solubilization, or solvent-phase solubilization. Depending on the type of contaminant to be cleaned and the chemical cleaner employed, one, two, or all three mechanisms are applicable.

Emulsification is the use of surface-active chemicals to colloidally disperse water-insoluble material into water. In other words, the insoluble material is broken down, temporarily suspended, and washed out of the machine clothing. The surfactant also reduces surface tension of the water, making it easier to flush through the wire or felt.

Water-phase solubilization refers to the process of solubilizing contaminants in acid or alkaline-treated water. Acid-soluble contaminants include calcium carbonate, iron hydroxide, and calcium phosphate. Fatty acid, rosin acid, lignin, synthetic size, and aluminum oxide are alkali-soluble.

Solvent-phase solubilization refers to the process of solubilizing contaminants in an organic solvent. Deposits such as asphalt, tar, latex, pitch, and wax are solubilized in solvent, emulsified with water, and removed. Surfactants are generally required for emulsification because of the limited water solubility of most solvents.

Papermachine clothing cleaners and conditioners are supplied in liquid form with active ingredient concentrations of 10-100 percent.

Regulatory Information

Any cleaner-conditioner must meet all local, state, and federal regulations. If the chemical is used in making food-grade related products, then every component of the cleaner-conditioner must be FDA-approved. Also, an MSDS must be provided with every cleaner-conditioner listing the hazardous ingredients, physical and chemical characteristics, fire and explosion data, physical hazards, health hazards, first-aid procedures, spill and leak procedures, and special protection information.

Criteria for Selection

Three steps are necessary to initiate an effective cleaning and conditioning program:

1. Identify and analyze the contaminants.

2. Develop a cleaning strategy.

3. Implement the program.

These three steps should always be followed by an on-line evaluation of the cleaning process.

Identifying contaminants is critical to selecting and developing the most effective cleaner. No two contaminants are exactly alike because the state of the contaminant is dependent upon furnish as well as paper machine conditions. However, an initial step in identifying a contaminant is to categorize it as either organic or inorganic.

Organic contaminants get trapped in the batt of the felt, thereby decreasing the porosity, absorption, and drainage of the felt. These contaminants can be a combination of any of the following:

1. Fatty and rosin acids
   Hydrocarbon oil and wax
   Triglyceride esters
   Calcium oxylate
   Lignosulfonate

2. Secondary fiber components
Adhesives
Hot melts
Plastics
Styrene butadiene copolymer inks
Carbon black

3. Treatment chemicals
Defoamer
Washer aid
Pitch control chemical
Fortified rosin size
Wet and dry strength aids
Retention aid
Synthetic size
Starches
Latex

A common way to differentiate between the alkali-soluble, acid-soluble, and solvent-soluble portions of the organic contaminant is by clothing extraction. In this process, samples of the contaminated clothing are subjected to hot alkali extraction, hot acid extraction, and room temperature organic extraction to determine the relative amounts of each type of contaminant. These extractions are performed on three identical "as received" samples.

Inorganic contaminants have a tendency to adhere to individual clothing fibers and remain on the surface yarns. These contaminants affect clothing flexibility, bulkiness, and resistance to abrasion. They can be a combination of the following:

Fillers (TiO_2, Clay, $CaCO_3$)
Coatings
Talc
Scale (insoluble salts, hydroxides, oxides).

The insoluble salts might be barium sulfate, calcium sulfate, calcium phospate, iron phospate, aluminum phospate, or magnesium silicate. The hydroxides and oxides can be calcium hydroxide, ferric hydroxide, or aluminum oxide.

In addition to identifying the nature of the wire or felt contaminants, the white water chemistry must be taken into consideration when developing an overall cleaning strategy. Only cleaner components that do not change the color of, or form a precipitate with, the white water are suitable for inclusion in a chemical cleaning solution. The nature of the contaminants, wet-end chemistry, personnel safety, FDA status, roll covering, yarn safety, and application equipment must all be considered when selecting an appropriate chemical detergent.

The cleaning action, or detergency, of clothing cleaners falls into one of three categories. First, the chemical bond between the wire or felt yarns and the "soil" can be broken by ion exchange, solvency, or more complex detergency. Second, a cleaner may disperse the contaminant particles so they flow out of the wire or felt with the water. The third type of cleaning action is a mechanical flushing of the wires or felts. The cleaner reduces the viscosity, or surface tension, of the water making it easier for the water to flow through the contaminated clothing.

Optimization of the selected clothing cleaner can only be achieved through proper implementation. Chemical reactivity, shower water temperature and pressure, and mechanical action must be balanced for optimum cleaner performance.

Storage and Handling Equipment

Cleaner and conditioning formulations can be shipped from the chemical manufacturing plant to the paper mill in Department of Transportaiton (DoT)-approved 55-gallon drums, 300-500 gallon tote bins, or bulk tankers. Selection of one of these alternatives depends upon the monthly usage, floor space near the machine, and availability of personnel to rotate drums or bins. A factor of increasing importance today is disposal of non-returnable drums and bins. Potential liability is so great that many mills are electing to prohibit the acceptance of any chemical container that would subsequently require disposal.

Tote bins are becoming increasingly popular in cases where mills use 50-100 gallons per day of a given product. A tote bin is frequently set up as a permanent day tank with appropriate piping and valving to feed metering pumps. A second tote bin set on top of the permanent bin can be equipped to feed directly into the latter bin. This provides the operator with storage capacity of 300-600 gallons.

In multi-machine mills where a single product is to be used for continuous wire and felt cleaning, or where batch cleaner products are used in large quantities, bulk storage is frequently the most efficient means of handling cleaner products. Tank sizes are frequently selected to accommodate 6,000-10,000 gallons of product. A pump located at the base of the storage tank is used to transfer the cleaner to a day tank set up near the paper machine. This pump can be activated by level controls on the day tank, by the control room operator viewing a display of day tank level, or manually by an operator. Tank construction varies with the type of cleaner being used. Mild steel, stainless steel, filament-wound fiberglass-reinforced plastics, and chemically-resistant polyolefins are examples of the types of materials encountered. Appropriate dike containment is recommended especially if the product has a high acidity or a high alkalinity. In many cases, dikes are required by regulatory agencies.

Drums, bins, and bulk storage tanks must carry labeling that conforms to the

requirements of all regulatory agencies such as DoT and OSHA. Products are rated according to health hazards, flammability, reactivity, and required personnel protection equipment. These ratings must be clearly displayed and are usually in a color-highlighted format.

In the section on application and use, some options for delivering water, cleaning solutions, and cleaning emulsions to the clothing surface will be described in detail. Some comments on delivering the concentrated product from the day tank to the line supplying the showers deserve attention. For both economic and papermaking process control reasons, it is essential that the quantity of cleaner be controlled at a constant level of addition. The units most frequently referred to in continuous cleaning are milliliters per minute. Units of gallons per minute are used in batch cleaning. For low flow rates, delivery is best accomplished by diaphragm pumps with chemically resistant internal wetted parts. These pumps can deliver up to 30-1,000 milliliters per minute against line pressures of up to 700 psig.

For optimum results, the pump should be set up so as to flood the pump's suction. It is extremely important to set up valve-isolated graduated drawdown cylinders on the suction side of the pumps. This is the only inexpensive, accurate way to establish and monitor chemical flow rates to the shower system.

The tubing piping from the pump discharge to the shower water line must be designed to handle the line pressure of the shower supply. Check valves and manual cut-off valves are highly-recommended at the nipple where the chemical feed line joins the shower water line. Much of this piping is on the crowded drive side of paper machines where piping failures can go undetected until runnability problems develop. Careful engineering and quality pipefitting will pay off in the long run.

A recent innovation, the patented pressure amplifier and condenser (PAC), has found widespread use in paper machine clothing cleaning. This unit converts mill cold water and steam into a pressurized flow of hot water. In the PAC chamber, sufficient vacuum is developed to pull 3-5 gallons per minute of concentrated chemical into the turbulent flow where complete emulsification occurs. The PAC replaces the heat exchanger, high-pressure water pumps, and chemical-feed pumps, making it an excellent component of cleaning systems. For continuous cleaning applications, chemical cleaners should be metered into the PAC by diaphragm pumps with antisyphon valves and check valves in the pump discharge piping.

Safety and Precautions

The following five subsections are meant only as general guidelines. All information regarding potential hazards and safety, specific to a cleaner-conditioner is detailed in the appropriate product MSDS.

Handling Precautions

Eye protection and gloves are usually required for safe handling of a cleaner-conditioner. Personal clothing should be laundered before use. Drums and tote bins should be stored in a cool place and tightly closed.

Protective Equipment

Every product is letter-coded to represent the level of personal protection required for adequate safety when working with the cleaner-conditioner. The codes represent personal protection ranging from safety glasses and gloves to full rain gear and respirators.

First Aid

Generally, eyes should be washed with large quantities of water for at least fifteen minutes, followed by medical attention. Skin should also be washed with large quantities of water. If inhalation occurs, remove the victim to fresh air, administer oxygen if necessary and seek medical attention. If a cleaner-conditioner is ingested, do not induce vomiting. Give water to cause dilution in stomach and seek medical attention.

Handling Spills

Contain the spill. Eliminate any ignition source. Absorb small spills with sand, dirt, or clay. Large spills may be pumped into containers for recovery or disposal in compliance with all applicable regulatory agencies.

Waste Materials Disposal

Consult federal, state, and local regulations. Contaminated absorbant may be deposited in a landfill, incinerated, or neutralized and flushed down the sewer.

Application and Use

Factors to consider regarding clothing cleaner application include the following:

1. Application equipment
       Type of nozzle (fan or needle)
       Feed equipment

2. Method of application
       Batch or continuous cleaning
       Application variables
       Shower water pressure
       Shower temperature
       pH of shower water

3. Application procedures

4. Treatment rates

5. Points of addition
    Location of showers
    Jet angle

The properly applied cleaner must then be monitored and evaluated.

Application Equipment

There are a few paper mills that clean forming fabrics strictly by mechanical means. One system features a two-brush cleaning apparatus in which the first rotating brush cleans the forming wire and the second rotating brush cleans the first. However, chemical cleaning and conditioning is the preferred method of maintaining machine clothing cleanliness. The cleaning solutions must cover the entire wire or felt surface on a regular basis. The solutions are applied via showers stretching the width of the machine. These showers may be fan or oscillating needle showers.

Fan showers generally run with a pressure of 40-100 psig and the nozzles are spaced so that the entire width of the wire or felt is covered as it passes over the shower. For single coverage, the spray pattern of one nozzle should just overlap the spray pattern of the next. A flooded-nip shower is a type of fan shower that is sometimes run on forming fabrics. A flooded-tip shower has the capacity to fill the running void volume in a forming fabric. Water is trapped in the nip and forced through the fabric at high velocity by the nip roll. This type of shower requires a high volume of water but is relatively safe with respect to fabric wear. A lube shower is a type of fan shower most commonly found on wet felts where it delivers a layer of water onto the surface of the felt, thereby providing lubrication between the felt and the uhle box.

Oscillating needle showers generally have a nozzle spacing of 3-12 inches on a shower that moves back and forth a distance equivalent to the nozzle spacing. There is no overlap of spray and the water pressure from each nozzle is generally 200-700 psig. An oscillating needle shower requires 10-20 oscillations to cover a wire or felt, but the high pressure tends to loosen and blow deposits out of the structure. Controlled speed and low-oscillation frequency are usually best. The oscillation must be uniform or it will cause uneven wetting and fabric wear problems. It is recommended that the oscillation frequency be such that the shower travels one-half the jet impact width, or less, per wire or felt revolution. Excessively high pressures can fibrillate individual yarns and "blow the nap away" and so should be avoided.

There are three designs of oscillating showers: pneumatic, hydraulic, and electro-mechanical. The electro-mechanical type is recommended for paper machines because of its low, adjustable oscillating speed.

A dryer felt needle shower that has received some recognition is a single nozzle transversing shower. A single high-pressure nozzle is followed by one or two air showers to dry the fabric. Potentially, this type of shower can be used to continuously clean dryer fabrics.

With any type of shower, the nozzles must be open and functioning as designed. Plugged nozzles will lead to streaks and profile problems. There is, however, a compromise to be made between nozzle orifice size and water conservation, spray angle, and impingement velocity.

Both fan and needle showers require a water and chemical feed system, as well as all the necessary input and output piping, valves, and gauges. A concept receiving a lot of recognition lately is the use of hot pressurized chemical solutions to clean clothing. This can easily be accomplished with cold mill water, steam, and a fabric cleaner via a device using steam to heat and pressurize the water and simultaneously mixing the cleaning chemical into the hot pressurized water. The hot pressurized cleaning solution is then fed through the showers to the machine clothing.

Application Methods

There are two primary methods of cleaning clothing, batch and continuous. Batch cleaning involves periodic cleaning with chemical concentrations of 5-20 percent. Since the sheet is taken off the machine, batch cleaning is by definition a downtime cleaning process. The advantages of batch cleaning are:

1. Generally less expensive than continuous cleaning.

2. No negative effects on paper properties.

3. Allows concentration of a chemical in a specific area.

The definition of batch cleaning has been expanded over the past couple of years and some mills are now batch cleaning while the sheet is on the machine. This is sometimes referred to as on-line batch cleaning or mini boil-outs. The chemical must be carefully chosen to ensure against negative effects on paper properties when the clothing is batch cleaned on the run.

Continuous cleaning is a preferred method in terms of clothing conditioning and uniformity because the clothing is cleaned from the time it is installed. Continuous cleaning keeps a wire or felt absorbant, porous, smooth, and resistant to plugging. The advantages of continuous cleaning are:

1. Longer clothing life and better performance.

2. No lost production from taking the sheet off the machine.

3. Less labor intensive than batch cleaning.

In reality, both batch and continuous cleaning are usually employed on a given machine. The frequency of batch cleanings is determined by machine runnability.

Application Variables

Shower water pressure and temperature are critical factors in the effectiveness of any clothing cleaner. Fan showers generally have a pressure range of 40-100 psig. However, needle showers can run at almost any pressure. Recommended pressure, in terms of optimum clothing runnability and life, is 200-300 psig. At this pressure, some soil-loosening takes place on the surface. At the same time, detergent is driven into the wire or felt structure where contaminants have been forced by hydraulic pressure over the suction boxes or in the press nip. Shower location is also critical to removal of contaminants and will be covered in a later section.

To make up for the lost effectiveness of running needle showers at relatively low pressures, the cleaning solution should be hot. The ideal temperature will vary with application but is generally 160-190° F (71-88°C). Hot water offers the following advantages:

1. Chemical reactions proceed more quickly.

2. Water has lower viscosity so it can penetrate the clothing easier.

3. More thorough rinse is achieved.

4. Dryer is protected from thermal shock during dryer felt cleaning.

The pH of shower water is also an important application variable. The safest, but not always most effective, way to run cleaning and conditioning showers is to match shower water pH with headbox pH. The pH range in which showers can be safely run on a given machine is determined by performing compatibility tests.

Application Procedures

Downtime batch cleaning is done with the sheet off the machine and the wire or felt rotation slowed down. The cleaner is usually applied through fan showers for 10-30 minutes. The cleaner is worked into the yarns by the action of the wire or felt against the rolls. The clothing must be thoroughly rinsed before the sheet is brought back across the machine.

On-line batch cleaning is done with the machine running full speed and the sheet on. Cleaner is generally applied through fan showers for 5-15 minutes. This slug of concentrated chemical will clean the wire or felt.

Dryer felt cleaning is different than wire or wet felt cleaning because the dryer section is so hot and the dryer cans, as well as the dryer felts, need to be cleaned. Chemical cleaning should be done on a downtime batch basis rather than continuously. The dryer felt is used as a carrier to transport cleaner to the dryer cans. Application should be made to the top felt first so that full advantage can be taken of the chemical detergency. The following procedure is generally applicable:

1. Run dryer felts, without sheet, at 250 -500 fpm.

2. Saturate felt with hot water, 180 - 190°F (82-88°C). If hot water is not available, the dryer cans must first be cooled to approximately 120°F(49°C).

3. Apply dilute cleaning solution through a full machine-width fan shower. Allow detergent to work into the mesh of the felt by running felt for

10-15 minutes after chemical has been applied.

4. Thoroughly rinse felt with hot water.

Treatment Rates

Continuous wet felt and wire cleaners are usually formulated to be used at 1-4 ml/gal of shower water or, 200-1000 ppm. The cleaner-conditioners are designed for shower flows of 1.5-2 gpm of water for each foot of machine width. Batch cleaners, whether for downtime cleaning or on-line cleaning of wires or wet felts, are used at a concentration of 5-20 percent. Dryer felts are used at a concentration of 3-6 percent.

Points of Addition

Cleaners are applied through machine-width showers. The preferred shower placement and type of shower depends on the clothing to be cleaned.

Needle showers or high-pressure oscillating fan showers are used for continuous cleaning on forming fabrics. These showers are best mounted on the outside of the wire, directly over a roll so that the roll acts as a backing to the shower jet. Low-pressure fan showers are used for batch cleaning. These showers are usually mounted on the inside of the wire near the the first inside return roll. Low-pressure showers are best mounted so that the shower spray hits the forming fabric on the ingoing side of a nip.

Wet felts are cleaned with oscillating needle showers or fan showers. The showers can be on the inside or the outside of the felt run. Inside showers are preferred when the felts are heavy, press nip pli is 1000+, press felt runs are long, and there is sufficient vacuum to work the detergent through the wet felts. Outside, or face-side cleaning is preferred when the felts are relatively light and the contamination is subject to rapid chemical interaction. In either case, the showers should be mounted to take advantage of the mechanical interaction of the felt with the rolls. Showers should be placed near the sheet release because the felts are more receptive to washing solution immediately after leaving the sheet. Also, impurities have less time to affix themselves to the felt and cleaners have more time to saturate and clean the felt. In reality, a face-side lube shower is often mounted about six inches before the uhle box and a face-side oscillating needle shower is mounted two to three feet before the lube shower.

Dryer felts are cleaned with a stationary fan shower at approximately 150 psig. The shower is placed so that the detergent has maximum dwell time in the felt. It is mounted at the highest point possible in the felt run to maximize the benefits of the drain-off of chemical rinse water.

Control Procedures

A number of machine parameters can be monitored to determine the effectiveness of a clothing cleaner (see Table 1). Some of these are on-line measurements, some are off-line, but all are important to objectively determine cleaner effectiveness.

Two pieces of equipment commonly used for on-line felt measurements are the Scanpro and the Huyck-Smyth Porosity Tester (HSPT). The Scanpro is used for moisture profiling wet felts. The HSPT is used to profile felt porosity.

Evaluation of Effectiveness

The ultimate evaluation of a chemical cleaning program effectiveness is clothing life, machine productivity, and product uniformity. Clothing life is the number of days a wire or felt runs on a machine. Machine productivity can be measured in a number of different ways but some of the

more common parameters are machine speed, tons produced, steam consumption per ton of production, and downtime. Product uniformity can be evaluated by cross direction moisture profiles and the absence of translucent spots.

Potential Problems and Solutions

Strong acid and alkaline cleaners must be used with prudence. Highly-alkaline cleaners can degrade dryer felt seams and may be damaging to polyester. Strong acids may damage nylon and can be corrosive to paper machines. Strong acid cleaners can also set rosin size into clothing rather than cleaning it out.

Cleaning chemicals, regardless of their makeup, should be checked for system compatibility. A chemical-system compatibility check is not difficult to run. Add 0.5 percent cleaner to a filtered wire pit sample, heat the solution to 160°F (71°C), and look for discoloration or precipitation. No change in appearance of the wire pit sample indicates compatibility between the cleaner and the paper machine system.

A disturbance of the machine balance, pH shock, results from adding a substance with a significantly different pH than that of the paper machine. In many cases, the fear of using a cleaner with a pH of 9 or 3 on a machine running a headbox pH of 5 is unwarranted because at cleaner use concentrations there is not much difference between cleaner pH and headbox pH. Also, many machines are buffered enough that minor changes in pH do not affect stability. Any incompatibilities caused by pH changes will surface when compatibility tests are evaluated.

Table 1.   Machine Parameters for Monitoring Clothing Cleaner Effectiveness

Monitoring Data:

- Style and design
- Running length
- Caliper
- Machine speed and grade
- Sheet basis weight and density profiles
- Headbox freeness
- Couch vacuum
- Press loadings
- Suction roll vacuums
- Suction box vacuums
- Shower water pressure and temperature
- Feed rate of cleaner/conditioner
- Wet felt moisture profiles before and after uhle boxes
- Uniformity of felt trade lines
- Condition of shower nozzles and oscillators
- Visual inspection of clothing
- Condition of uhle box cover
- Deposits around uhle box cover
- Deposits on wire and felt rolls, return rolls
- Vacuum pump separator flow
- Air flow through forming fabrics
- Clothing filling at time of removal

Sources of Additional Information

1. Houfek, W. E., "Simple Maintenance Procedures Can Help Extend Forming Fabric Life," *Pulp Paper* 62(4):163-164 (1988).

2. Green, M. R., "Showers and Doctors on the Twin-Wire Former," TAPPI 1986 Twin Wire Seminar (New Orleans), Note: 87-97 (1986).

3. Treece, R. A.,"Useful Life of Dryer Felts is Extended with Scrupulous Cleaning," *Pulp Paper* 60(2):69-70, 74 (1986).

4. Furibondo, N., "Wire, Felt Cleaning Programs Can Improve Paper Machine Runnability," *Pulp Paper* 61(8):119-122 (1987).

5. Stevens, O. C., "Dryer Felt Designs Complement Modern Paper Machine Systems," *Paper Trade J.*69(8):45-47 (1985).

6. Turner, G., "Cleaning Secondary Fiber from Dryers and Dryer Felts," *Southern Pulp Paper Manufacturer* 48(3):16,18 (1985).

7. Shyamsunder, R., "Machine Clothing [for] Mysore [Paper Mills Ltd.] Newsprint Machine," *IPPTA* 20(4):86-93 (1983).

8. Daral, R. N., and A. G. Alate, "Chemical Conditioning of Wet-Section Felts: Why is it Necessary?," *IPPTA* 20(4):61-65 (1983).

9. Green, M. R., "Low-Frequency Oscillators Used for Felt and Fabric Cleaning Showers," CPPA 1986 Annual Meeting Preprints 72A:225-227 (1986).

10. Yokoyama, K., "Countermeasures Against Contaminations of Plastic Wire and Felts," *Japan Pulp Paper* 29(3):71-76 (1985).

11. Olsson, J. L., "Study Shows Greater Potential for High-Pressure Showers," *Paper Technology Ind.* 27(1):34-36, 38 (1986).

12. Reese, R. A., "Anticipating Press Felt Performance Problems," *Tappi J.* 66(10):11 (1983).

13. Fendrick, E., "Cleaning and Care of [Papermaking] Felts," *ATIPCA* 21(2):34-35 (1987).

14. Shew, C. F., "How A Papermaker Improved on Cleaning Paper Machine Fabrics," *PIMA* 69(2):34-35 (1987).

15. Meinecke, A., and J. M. Voith GmbH., *Apparatus for Cleaning Papermaking Machine Screen Belts.*

16. Ericson, L., "Wire Cleaning by FAW-Brush," KMW/Scandia Felt Tissue Making Seminar (Karlstad): 1985, p. 10.

17. Nishimuta, J., "Press Section Optimization - Monitoring of Wet Press Felts," 1990 Papermaker's Conference (Atlanta), Note: 11-14 (1990).

Chapter 4

Alum

by Barbara H. Wortley

Introduction

Alum (aluminum sulfate, with the formula $Al_2(SO_4)_3 \cdot nH_2O$ where "n" is approximately 14) has been widely used by papermakers for over 150 years to obtain one or several of the following benefits:

- Water purification
- pH adjustment
- Sizing
    Rosin
    Wax
- Retention
    Fines
    Filler
    Dye
    Starch
    Latex
    Dry-strength additive
- Drainage aid
- Pitch control
- Foam control
- Press picking reduction
- Wet-strength resin curing
- Wet-strength broke recovery
- Improved saveall operation
- Effluent treatment

Purpose of Use

The strong cationic charge of the polyvalent aluminum species is employed in most of the previously-mentioned applications. Alum is an integral part of rosin sizing systems and performs several important functions. When using soap rosin size, alum provides the acid condition and the aluminum ions which precipitate the rosin. In addition, alum provides a positive charge to the precipitate so that it can be deposited on the negatively charged fiber surface. Additionally, alum anchors the size precipitate during drying, causing the hydrophobic portion to remain properly oriented.

Different mechanisms control sizing development when using dispersed rosin acid sizes. There is little reaction between rosin and alum in the wet end. The interaction between rosin and hydrolyzed aluminum (which has adsorbed on the fiber) takes place during drying. The adsorbed aluminum provides a cationic patch for rosin attachment, while properly orienting and

Table 1. Product Description - Alum	
Aluminum Sulfate, Dry	
Chemical formula:	$Al_2(SO_4)_3 \cdot 14H_2O$ (approx.)
Molecular weight:	594 (approx.)
Appearance:	White to cream color
Solubility in H_2O:	Theoretical 50%, Practical 25%
pH of 1% solution:	3.5 (approx.)
Aluminum Sulfate, Liquid	
Formula:	Approximately 48.5% dry aluminum sulfate in water
Appearance:	Light green to light yellow liquid
Specific gravity, 60°F:	1.335
pH of 1% solution:	3.5 (approx.)
Baume', 60°F:	36.4
lb/gal (U.S.):	11.1
lb/gal (Imperial):	13.3
lb alum/gal (U.S.):	5.4
lb alum/gal (Imperial):	6.5

holding the hydrophobic portion of the size on the outer surface of the fiber.

Natural wood resins (pitch) and other organic contaminants such as latex and adhesives (white pitch) are also deposited on the fiber surface through interaction with alum. The pitch is then carried out with the sheet, preventing troublesome accumulation in the papermaking system.

Fines are retained in a sheet through physical entrapment and colloidal phenomena. Most fibers, fiber fragments, and mineral fillers (clay and TiO_2) are negatively charged in water and repel each other. The electrostatic mechanism of opposite charge attraction brings together the negatively charged material and the positively charged aluminum species.

Since ionic strength is based upon the square of ionic valence, very strong ionic interaction between the polyvalent aluminum species, fibers, and fines can result. Adsorption of aluminum on the fiber surface provides a cationic patch which attracts and holds negatively charged fines or fillers.

Aluminum addition plays a major role in enhancing drainage by neutralizing the usual anionic surface charge of fibers. Drainage rate increases as the isoelectric point (zero charge) is approached. In addition, deposition of fines onto fiber surfaces opens the sheet so that water can drain more freely.

Alum is also extensively used as a coagulant-flocculant for water and wastewater treatment.

Table 2. Typical Alum Analyses

Aluminum Sulfate, Dry

Grade	Commercial	Iron-free
Total Al_2O_3 (%)	17.1	17.2
Free Al_2O_3 (%)	0.2	0.3
Total iron as Fe_2O_3 (%)	0.3	0.005
Actual Fe_2O_3 (%)	0.05	---
Insoluble in water (%)	0.1	0.015

Aluminum Sulfate, Liquid

Grade	Commercial	Iron-free
Total Al_2O_3 (%)	8.3	8.3
Free Al_2O_3 (%)	0.1	0.1
Total iron as Fe_2O_3 (%)	0.2	0.004
Actual Fe_2O_3 (%)	0.03	---
Insoluble in water (%)	0.01	0.004

The dry material (powdered or granular) is generally available in multi-wall bags or bulk shipment. The liquid material is shipped in tank transport and tank cars.

The described mechanisms have been extensively covered in the literature. Consult Casey (1) and *TAPPI Monograph Series No. 33* (2) for more complete coverage. Davison (3) presents a review of alum application in the paper industry along with an extensive bibliography in *TAPPI CA Report No. 57*. Strazdins (4) (5) and Marton and Marton (6) (7) (8) also provide excellent reviews on uses of alum.

General Composition

Alum is supplied in dry or liquid form in commercial, iron-free, or food grades. Table 1 details typical product description. Table 2 details product analyses. However, products may vary and the particular product specifications must be obtained.

The dry product (powdered or granular) is generally available in multi-wall bags or bulk shipment. The liquid product is shipped in tank transport and tank cars.

Unless excessive shipping distances are involved, liquid product is more economical, on the basis of aluminum content, than dry product because it is an intermediate step in production. The evaporation, grinding, and bagging of the finished dry product add considerably to its production cost. If sufficient volume is consumed to warrant erection of storage facilities, using liquid product greatly simplifies operations.

Small amounts of iron adsorption may be detrimental to the brightness of high-yield pulps (groundwood, TMP) or interfere in blueprint paper production. Iron-free or low-iron alum may be the product of choice in such cases.

Storage and Handling Equipment
Shipping Containers

Dry alum (powdered, granular, or lump) is generally available in multi-wall bags or bulk shipment.

Liquid alum is shipped in stainless steel tank transports of up to 5,000 gallon capacities. Liquid alum transports are usually self-unloading by means of a pump or air compressor mounted on the tractor. If adequate facilities for handling tank cars are available, liquid alum can be received in rubber-lined tank cars of 10,000 and 20,000 gallon capacities. Unloading hoses and a source of compressed air not to exceed 30 psig must be provided by the receiver.

Bulk Storage Equipment

Fiberglass-reinforced polyester and other suitable plastic materials can be used for liquid alum storage tanks which may be of vertical or horizontal configuration. Tanks should be covered to prevent evaporation and contamination and must be properly vented. Storage tank capacity should be 1.5 times delivery vehicle capacity or more, depending upon use rate.

Greater physical strength for high-traffic areas or other special requirements may be provided by rubber-lined steel, type 316 stainless steel, or PVC bag liners in steel or wood stave tanks.

Storage Temperature

Liquid alum freezes at approximately 5°F (-15°C) and, under certain conditions such as surface evaporation, long-term storage, etc., crystallization may occur near 30°F (-1.1°C). Storage tanks should be installed indoors or in a heated enclosure. Storage tanks and equipment located outdoors should be heated with 5 psig steam, hot water in tubing, or electrical heating cable or coils, then insulated to maintain a temperature of 45-60°F (7-16°C).

Lines exposed to low temperatures should be insulated. Good engineering practice provides flush-out connections at strategic points in the piping, particularly in the area of pumps and metering equipment. These connections can then be used to purge the pipelines to an environmentally acceptable container with a minimum of effort should crystallization occur. Water lines used for flushing should be disconnected after use to minimize the possibility of contamination of water supply.

Feeding Equipment

Pressurizing a storage tank with air or nitrogen is not recommended as a means of transferring liquid alum. In addition, plastic storage tanks, frequently used for alum, cannot be pressurized.

Where gravity flow to process or feed equipment is practical and desirable, a savings in pump capital and maintenance costs is possible. If a gravity system is not possible, centrifugal or metering pumps, sized for head and flow requirements, may be used to transfer alum to the process.

Centrifugal pumps with wetted parts of Alloy 20, type 316 stainless steel, or plastic, sized for the total head and required flow, are recommended for transferring liquid alum from storage to process. Motor speeds over

1800 rpm should be avoided. Pumps operating at high speeds will have short lives and will require more maintenance.

Diaphragm-metering pumps with all wetted parts of Alloy 20 or type 316 stainless steel and TFE diaphragms are recommended. Magnetic flowmeters or rotometers of PVC or 316 stainless steel with Hastelloy C floats may be used. In-line strainers are recommended.

Schedule 80 PVC normal impact or chlorinated polyvinyl chloride pipe (CPVC) with solvent-weld socket fittings are recommended. PVC and CPVC, due to their high coefficient of expansion with temperature changes, should not be anchored at both ends of a piping run and are best supported continuously, such as in a channel or angle iron support. Schedule 10 or Schedule 40 type 316 stainless steel pipe with all welded fittings may be used.

Safety and Precautions

Though aluminum sulfate is not considered a particularly hazardous material, its buffered acidic action can be irritating when in contact with eyes, skin, or mucous membranes. This can be avoided with the use of protective clothing and equipment.

Protective Equipment

Normal precautions should be employed to prevent spraying or splashing liquid alum, particularly when it is hot. Face shields or eye goggles should be worn to protect eyes. Outer covering, such as rubber aprons and waterproof sleeves, may be used to protect clothing from liquid alum.

First-Aid Measures

Eyes. Flush immediately with water for at least 15 minutes. If irritation persists, seek medical attention.

Skin. Flush with plenty of soap and water. Remove contaminated clothing. If irritation develops, seek medical attention.

Inhalation. Promptly remove person to fresh air.

Ingestion. If conscious, immediately give a large quantity of water or milk. If not already vomiting, induce vomiting by touching finger to back of throat. Get medical evaluation.

Pollution Control Considerations

Accidental discharges of alum in substantial quantities can be harmful to the environment. It is recommended that outdoor storage tanks be suitably diked or otherwise provided with an adequate means of secondary containment.

Appropriate secondary containment measures should also be taken to prevent spills or leaks originating in indoor storage tanks, tank cars, or tank transport unloading stations from entering sewers or other channels that discharge directly to a body of water or a municipal sewage system.

Planned containment measures should be reviewed with appropriate water pollution control regulatory agencies to ensure proposed installation compliance with applicable laws and regulations (see Table 3).

Waste Disposal Methods

All waste disposal must comply with federal, state, and local disposal or discharge laws. Waste should be neutralized with alkali, then flushed to a sewer with plenty of water if permitted by applicable regulations.

Table 3. Regulatory Information - Alum

	Alum (Dry)	Liquid Alum	Reference
Trade Name	Alum (Dry)	Liquid Alum	Reference
Chemical Name	Aluminum Sulfate	Aluminum Sulfate, aqueous solution	
Formula	$Al_2(SO_4)_3 \cdot 14H_2O$ (approximately)	48.5% $Al_2(SO_4)_3 \cdot 14H_2O$ in water	
C.A.S. No.	10043-01-3 (anhydrous)	10043-01-3 (anhydrous)	
EPA Hazardous Substance (For Spill)	yes	yes	40 CFR 116-117
(Reportable Qty)	8,700 lb*	18,000 lb	
RCRA Status of Unused material	Not a Hazardous Waste	EPA Hazardous Waste No. D002 (corrosive) if pH $\leq$ 2	40 CFR 261
DoT Classification	ORM-E*	ORM-B	
DoT ID No.	NA 9188	NA 1760	49 CFR 172

* Not regulated if less than 8,700 lbs in one package.

Neutralized waste may require disposal by an approved contractor. When using carbonates for neutralization, adequate precautions should be taken to minimize hazards and pressure buildup from CO_2 gas evolution.

Application and Use

While the vast majority of the alum consumed by the paper industry is now purchased in liquid form, chemical addition and pricing are based on the dry product $Al_2(SO_4)_3 \cdot 14H_2O$. An alum addition of one percent refers to 20 pounds of $Al_2(SO_4)_3 \cdot 14H_2O$ per dry ton of pulp, (10 kg/mt). When applied as standard commercial liquid alum, this can be added volumetrically using the following calculation:

20 lb alum $\div$ 5.4 lb dry alum per gallon =

3.7 gal liquid alum per oven-dry ton of furnish (15.4L/metric ton).

As a carry-over from the days when alum was received in dry form and made down to the 1-2 pounds per gallon level (0.12-0.24 kg/L), some mills continue to dilute purchased liquid alum to the same concentration. Fortunately, the trend now leans toward using liquid alum as received. This change eliminates one process step, added equipment, and possible problems.

Scale formed from the reaction of alum with water hardness can plug lines. The dilution water can also introduce slime-forming microorganisms into the dilute alum storage system. Accurate metering equipment is available for feeding the concentrated liquid alum to process.

Treatment rates vary considerably depending upon grade produced, furnish, alum function, (pitch control, fines retention, sizing). Feed rates and points of addition are best determined empirically. Some compromises may be required. The ideal pH for maximizing fines retention may not be best for sizing development, etc.

The very complex chemistry of aluminum will not be covered in this discussion. Previous references and Arnson (9) provide additional detailed information. The electrostatic mechanism of opposite charge attraction is a very important function. Brief, simplified descriptions of the mechanisms encountered in the principal categories of alum application follow.

Fines Retention

Fibers, fines, and fillers are generally negatively charged when dispersed in water. Adsorption of positively charged aluminum on fibers forming cationic patches (positive charge) where negatively charged anionic fines and fillers can attach themselves is a very important aspect of fines retention. Aluminum adsorption is a function of system pH and aluminum concentration. Aluminum is not readily adsorbed by cellulose at a low pH (<4.5). A rapid rise in aluminum adsorption has been shown (9) (10) to coincide with the formation of the highly charged polynuclear aluminum specie, proposed to be $Al_8(OH)_{10}(SO_4)_5^{+4}$. The actual polynuclear species which can form have not been unequivocally identified.

There appears to be more evidence of the formation of the Al_{13} specie. Generally, operating in the range of pH 4.8-5.2 will greatly enhance aluminum adsorption and, therefore, fines retention. The critical pH for polynuclear formation will vary with each system. Both pH and alum concentration must be optimized and controlled independently of the other.

Sizing

The aluminum/rosinate reaction is the key function in alum/rosin sizing (11). The alum:rosin ratio is generally 1.5-2:1. Alum reacts rapidly in solution with rosin soap size (paste rosin) to form the insoluble aluminum-rosinate complex, which has a strong cationic charge due to the aluminum on the surface. This charge mechanism attracts the rather voluminous precipitate to the negatively charged fiber. Alternately, with a dispersed (emulsified) rosin acid size there is only limited reaction in the wet end between the alum and the surface of the dispersed rosin particle. Under proper pH conditions, aluminum is preferentially adsorbed by the cellulose. The acid size reacts with the preadsorbed aluminum during the drying operation.

The addition sequence, point of addition, pH, and concentration of size and alum must be carefully considered (along with the many other substances present in the papermaking system which can add or detract from sizing efficiency) when optimizing a system. Good, rapid mixing when adding the size and the alum is very important.

When using a dispersed acid size, a simple approach would be to operate at pH 4.8-5.5 which favors the formation of hydrolyzed aluminum species and increased aluminum adsorption. A lower pH system in which Al^{+3} predominates would be more acceptable for soap size. Reverse sizing (alum followed by size) in the pH range favoring aluminum adsorption by the cellulose is generally avoided when using soluble soap size. The required soluble aluminum may not be present and the size may tend to agglomerate before attachment to the fibers. Soap size generally contains some free resin acid and, conversely, dispersed size will have some soluble size present thereby adding a degree of variability. At times (due to hard water, etc.) better results are obtained by operating in the reverse mode, even with a soap size. Each paper machine system is unique, and a careful study of the many variables and interactions must be made.

Pitch Control

Alum is used to precipitate natural wood resins (pitch) onto fibers using mechanisms similar to those encountered in alum/rosin sizing. Pitch precipitated onto fiber is carried out of the system with the paper, preventing troublesome buildup within the system. Alum used for groundwood pitch control is frequently added directly to the grinder pit. Additional alum may be applied prior to sheet formation to capture any pitch released as a result of mechanical action on the fibers.

Alum Addition Control

Because alum traditionally has been used to add acidity to the papermaking system, pH has long been used to control alum addition. The buffering effect which occurs in an alum feed system can result in excessive alum feeds when controlling alum addition by pH alone. Regulating alum addition by maintaining a specified total acidity provides a more meaningful control. Total acidity indicates the soluble (unneutralized) alum

available in the system. Optimal total acidity for a particular system must be determined empirically. A total acidity reading of approximately 50-80 ppm in the tray water generally ensures adequate soluble aluminum is available when using soap sizes while avoiding detrimental excesses. Little or no soluble alum is required for proper application of dispersed acid size.

Much more significant information can be obtained by measuring the actual aluminum content at various points in the system. The determination of alum in papermaking furnishes has been greatly facilitated through the development of the specific fluoride ion electrode (SFIE) method. Avery (12) describes the procedure and its application in laboratory and mill settings.

Evaluation of Effectiveness

A quick assessment of fines retention can be made by comparing the solids content of the traywater with that of the headbox. Excessive buildup of fines in the headbox due to poor first-pass retention can cause many problems such as two-sidedness, reduced sizing efficiency, press picking, linting, etc. Laboratory evaluations can be made using a dynamic drainage jar.

There are many techniques for evaluating the degree of sizing development. Refer to Chapter 6 of this guide and to TAPPI Standard and Useful Test Methods.

Potential Problems

Commercial liquid alum (48.5 percent solution) is at its optimum concentration for shipping and storage purposes. The freezing point rises dramatically as the product concentration approaches 50 percent. Care should be taken to prevent concentration through evaporation or by the addition of other chemicals.

Insoluble aluminum hydroxide, $Al(OH)_3$, forms at elevated pH. Shower water pH should be kept close to the headbox pH to prevent aluminum deposits on the wire, dandy, couch roll, or felts. Slime and pipeline deposits caused by process water contamination when diluting alum have been covered previously.

Literature Citations

1. Casey, J. P., *Pulp and Paper*, 3rd edn., New York: John Wiley & Sons, Vol. 3, p. 1562-1565.

2. Reynolds, W. F., ed., *The Sizing of Paper*, 2nd edn., TAPPI PRESS, 1989, p. 1.

3. Davison, R.W., "Retention of Fine Solids During Paper Manufacture," *TAPPI CA Report No. 57*, Atlanta, GA: TAPPI PRESS, 1975, p. 71.

4. Strazdins, E., *Tappi J.* 64(1):31 (1981).

5. Strazdins, E., *Tappi J.* 69(4):111 (1986).

6. Marton, J., and T. Marton, "Some New Principles to Optimize Rosin Sizing," *TAPPI 1982 Papermaker's Conference Proceedings*, Atlanta, GA: TAPPI PRESS, 1982, p. 15.

7. Marton, J., and T. Marton, *Pulp & Paper Canada* 83(11):T 303 (1982).

8. Marton, J., and F. L. Kurrle, "Retention of Rosin Size," *TAPPI 1985 Papermaker's Conference Proceedings*, Atlanta, GA: TAPPI PRESS, 1985, p. 197.

9. Arnson, T. R., *TAPPI 1981 Sizing Seminar Course Notes*, Atlanta, GA: TAPPI PRESS, 1981, p. 41.

10. Wortley, B. H., *TAPPI 1987 Advanced Topics in Wet-End Chemistry Seminar Course Notes*, Atlanta, GA: TAPPI PRESS, 1987, p.55.

11. Wortley, B. H., *PIMA Magazine* 70(2):51 (1988).

12. Avery, L. P., *TAPPI 1978 Retention and Drainage Technology in Paper Manufacturing Seminar Course Notes*, Atlanta, GA: TAPPI PRESS, 1978.

Chapter 5

Internal Sizing Agents

by Robert E. Cates

Introduction

Internal sizing agents are one of the most important additives used in paper. Most grades of paper and paperboard, except absorbent grades, require internal sizing agents. The sizing may be necessary for the end use of the paper, as in milk cartons, linerboard, or bag papers. Alternatively, sizing may be required only for operability on the paper machine, to give holdout of size press, calender, or coating applications.

Conventional internal sizing agents impart resistance to water and aqueous liquids only. They do not produce sizing to oils, solvents, or oil-based inks. Oil resistance may be obtained by the application of oil-resistant films, or, internally by the use of fluorochemicals (1).

An effective internal sizing agent must meet the following requirements:

1. It must have the ability to be retained on the cellulose fiber.

2. It must be uniformly distributed on the fiber.

3. It must have good adhesion to the fiber.

4. It must be naturally hydrophobic or reach a hydrophobic condition during the drying operation.

It is beyond the scope of this chapter to discuss sizing theory or sizing mechanisms. The reader is referred to *TAPPI Monograph Series No. 33*, "Internal Sizing of Paper and Paperboard," (2), *Sizing of Paper*, 2nd edn. (3) or *Pulp and Paper Chemistry and Chemical Technology* by James P. Casey (4).

For many years, the available internal sizing agents operated using alum at acidic wet-end pH. The presence of high levels of alum, and buffering of the paper at the pH 4 level, significantly reduces the strength of the finished paper. In recent years, new cellulose-reactive sizes have been developed to operate at neutral to alkaline pH, without alum, or at very low alum levels. Sizing agents for both acid systems and neutral-to-alkaline systems will be discussed in this chapter.

Rosin Size
Types of Rosin Size

Rosin is a naturally occuring resinous material obtained from the pine tree. The major constituent of rosin is a family of tricyclic organic acids called resin acids. The aluminum salt of rosin is water-repellent and has been used for the sizing of paper since the earliest days of continuous papermaking.

Saponified rosin size is made by reacting either unmodified or chemically modified rosin with an alkali such as sodium hydroxide. The resulting rosin soap is water-soluble and can be mixed with paper stock at the wet end of the paper machine. When precipitated by papermakers' alum, aluminum sulfate, the resulting aluminum resinate becomes cationic for retention on the fiber and develops water repellency upon drying.

Dispersed rosin size, also called rosin emulsion size, is produced by emulsifying rosin in water with little or no saponification of the resin acids. This high-free rosin size is capable of being retained on the cellulose fiber by alum, or suitable cationic retention aids, and reacts with alum in the dryer section to produce a hydrophobic mixture of free rosin and aluminum resinate on the fiber surfaces.

Saponified rosin size and dispersed rosin size are completely different in physical properties, handling, and use, and each will be discussed separately.

Saponified Rosin Size

Saponified rosin size prepared from unmodified rosin is known as unfortified rosin size. When the rosin has been chemically modified by reaction with maleic anhydride or fumaric acid, the resulting saponified size is known as fortified rosin size. Fortified rosin size is more cost-effective in bleached pulps, while unfortified size is used primarily in unbleached kraft grades.

Saponified rosin size is available in pale and dark grades. Pale size is prepared from the lighter-colored grades of rosin and is suitable for use in high-brightness paper and paperboard. Dark rosin size is prepared from more highly colored grades of rosin and is used in low-brightness or unbleached grades of paper and board. Saponified rosin size is available in paste, fluid, and dry forms.

Paste Rosin Size

Paste rosin size, which is 70-77 percent solids, is shipped in tank trucks, tank cars, or in standard steel drums. It is semi-solid at room temperature and must be heated to about 180°F (82°C) for handling. Typical properties of paste rosin size are detailed in Table 1.

Table 1. Typical Properties of Paste Rosin Size

	70% solids	77% Solids
Total solids, weight %	70	77
Appearance	Paste to viscous liquid	Paste to viscous liquid
Density at 180°F, lbs/gal	9.3	9.5
Viscosity at 180°F, cps	400	1000
pH (dilute emulsion)	9-10	9-10
Effect of freezing	Minimal. Thaw and mix before using.	Minimal. Thaw and mix before using.
Shelf life	1 year. Avoid heating for prolonged periods.	1 year. Avoid heating for prolonged periods.

Shipment of Paste Rosin Size

When shipped in tank trucks, paste rosin size is shipped by the manufacturer at 180-200°F (82-93°C) and remains hot enough for unloading at the destination without reheating. Unloading may occur by air padding of the tank truck, use of a pump located on the truck, or use of a pump located at the unloading site. A 35 psig steam connection with flexible steam hose must be available at the unloading site for preheating of the unloading line and blowing out the size line after unloading.

Tank cars of paste rosin size become cool in transit and must always be heated before unloading. All paste size tank cars are equipped with interior coils, or with steam jackets on the exterior of the tank, for heating. All paste size tank cars are insulated.

Heating and Unloading Paste Rosin Size

For heating paste rosin size in tank cars, steam pressure should be 35 psig maximum. Higher steam pressure, and resulting high heating surface temperature, can bake the size onto the heating surface, greatly reducing heat transfer. The dessicated size eventually breaks away from the heating surfaces and may appear as flakes in the size.

Paste rosin size must be heated to a minimum temperature of 180°F (82°C), measured at the top of the tank car, before unloading. Major size suppliers can provide detailed written procedures for unloading of size tank cars.

The paste rosin size unloading line should never be less than three inches in diameter and should be steam-traced and insulated. For long distances or highly elevated unloading lines, a four-inch line is recommended.

Paste Rosin Size Storage Tank Design

Since paste size is noncorrosive, the paste rosin size bulk storage tank is usually made of mild steel. However, tile or stainless steel is recommended due to eventual iron scale formation in the vapor space of a mild steel tank, especially in a corrosive atmosphere, such as adjacent to a bleach plant.

The paste rosin size storage tank must be insulated and equipped with a heating compartment over the outlet to keep the size warm enough to flow to emulsifying equipment. Heating of the entire tank contents is not recommended due to danger of evaporating water from the surface, causing formation of a crust which will not redissolve in the size. Energy costs for handling paste size are substantially reduced when only the portion immediately over the outlet is kept heated. A temperature of 180°F (82°C)in the heating compartment is usually recommended.

For heating calculations involving paste rosin size, the following heat transfer coefficients are used:

5 BTU/hour/ft^2/°F for 77 percent size

8 BTU/hour/ft^2/°F for 70 percent size

These coefficients provide reserve heating capacity for adverse conditions, cold start-ups, etc. The specific heat for all sizes is 0.65 BTU/lb/°F. Steam pressure for heating paste rosin size is 35 psig maximum for the reasons noted earlier. Specifications recommending designs and required heating areas for paste size heating systems are available from major size manufacturers.

Emulsification of Paste Rosin Size

Paste rosin size is a mixture of rosin soap and unsaponified rosin. Because of its high viscosity and the fact that it is not directly dilutable with cold water, paste size must be emulsified to a 2-5 percent concentration before addition to the papermaking system. Cold, fresh water with hardness below 100 ppm is required for emulsification.

A primary emulsion is first prepared by mixing 180°F (82°C) size with 180°F hot water at 12-16 percent size concentration. This is then diluted to 2-5 percent solids with cold water to give a final emulsion at 60-100°F (16-38°C). Emulsification may be carried out in manually operated batch equipment, or in special automatic equipment.

Handling and Metering of Rosin Size Emulsion

Rosin size diluted to 2-5 percent solids has the essential properties of water which should be used for calculating pump sizes and pipe line sizes. Dilute size emulsion is noncorrosive and can be handled in black iron piping. Plastic, plastic-lined, or stainless steel pipes are also used. Control of size emulsion flow to the paper stock is usually by rotometers with manual valves, or by magnetic flowmeters with automatic valves. Rosin size emulsion will gradually coat metering equipment surfaces. Rotometers or magnetic flowmeters should be installed in an offset to the pipeline to provide easy access for regular cleaning. Due to the tendency of deposits to build up in emulsion tanks and pipelines, a 20-40 mesh final strainer or filter is recommended at the point of use.

Use of Paste Rosin Size in Sizing Paper

The usual method of rosin size application is to add the dilute size emulsion to pulp that has been adjusted to pH 7-7.5. The dilute size may be added batchwise at the beater or pulper, or continuously to the pulp at the machine chest or stuffbox. Alum is added after the size to a final headbox pH of about 4.2-4.7. A minimum of 1 percent alum, based on the fiber, is recommended.

Under some adverse sizing conditions, such as high hardness, alkalinity, or interfering ions in the stock, it is preferable to add the alum first to ensure the rosin size reacts with alum rather than with hardness or other interfering ions. This is known as reversed sizing procedure.

The reaction of rosin size with alum produces a size-alum precipitate, which is cationic and retained on the anionic pulp fibers by electrostatic attraction. The size-alum precipitate reorients on the fiber during drying to develop a water-repellent fiber surface.

Paste rosin size is usually used at levels of about 0.1-1.25 percent, dry solids basis, depending on the sizing requirements of the paper. The control of wet end conditions for saponified rosin size is more critical than for most other paper additives because the actual sizing material, aluminum resinate, is manufactured in situ in the pulp. The composition and properties of the aluminum resinate precipitate are very dependent upon wet end conditions. Because the requirements of different pulps, different grades of paper, and different papermaking equipment vary widely, major size manufacturers provide technical service in using and controlling their products. The paper manufacturer should take advantage of this technical service since poor control of rosin sizing conditions may not only result in high sizing costs, but also result in operating problems such as foam, press picking, and spots in the paper.

Fluid Rosin Size

Fluid rosin size differs from paste rosin size in three respects:

1. The solids are 35-60 percent, instead of 70-77 percent.

Table 2. Typical Properties of Fluid Rosin Size	
Total solids, weight %	35-60
Appearance	Water-thin to syrupy liquid
Viscosity at 77°F, cps	3-1700
Density at 77°F, lbs/gal	9.3-9.8
pH	10
Effect of freezing	Minimal. Thaw and stir before using.
Shelf life	1 year. May stratify during prolonged storage.

2. The size is fluid at room temperature and requires no heating for handling.

3. The size is completely saponified and is soluble directly in paper stock or in cold water without an emulsification step.

Typical properties of fluid rosin size are shown in Table 2. Fluid size is usually more costly than paste size on a delivered active solids basis. Fluid size is usually used where size usage is low and variable, or when adequate heated storage or proper emulsifying equipment for paste size is not available.

Storage of Fluid Rosin Size

As with paste rosin size, bulk storage tanks for fluid rosin size may be made of mild steel and pipelines may be made of black iron. Tile, stainless steel, and plastic-lined tanks and pipelines are also suitable. In mills using small quantities of size, fluid size is purchased and stored in steel drums.

Diluting Fluid Rosin Size

Fluid sizes may be diluted in batch or automatic dilution equipment. The same equipment used for emulsifying paste rosin size can be used, except that heating of the size and of a portion of the water is not required. Dilute fluid size solution has the essential properties of water which should be used in calculating pump sizes or pipeline sizes.

Metering and Using Fluid Rosin Size

Fluid size is often metered in concentrated form to the point of addition, using suitable pumps and metering equipment. After metering, the concentrated size is diluted with a stream of fresh water to about 2-5 percent solids before addition to the pulp.

All recommendations for metering dilute paste size emulsion and using it with alum in paper sizing as outlined earlier also apply to the use of fluid size.

Dry Rosin Size

Dry rosin size is a completely saponified size that has been dried to a free-flowing powder. It is usually shipped in multi-wall paper bags containing 50 pounds net. Typical properties of dry rosin size are shown in Table 3.

Due to the energy costs of removing water, dry rosin size is higher in price per dry pound than a corresponding grade of paste size. This disadvantage may be recovered in lower freight costs, especially in remote locations or in foreign countries. Dry size is usually used where rosin size usage is small and highly variable, or when suitable storage and handling equipment for paste size is not available. Some paste size users keep dry size on hand for emergency use when paste size emulsifying or metering equipment malfunctions.

Although dry rosin size is shipped in bags containing a vapor barrier, the hygroscopic nature of dry size causes moisture to enter the bag gradually through the bag ends. For this reason, dry rosin size must be stored in a cool dry place and stocks should be rotated at least once a year to prevent caking and lumping.

Handling and Using Dry Rosin Size

Dry rosin size is cold water-soluble and can be added directly to preneutralized paper stock. It may also be dissolved in cold or warm water to give a 2-5 percent solution which may be added batchwise or continuously to the pulp.

Table 3. Properties of Dry Rosin Size

Total solids, weight %	9 6
Appearance	Free-flowing powder
pH (dilute solution)	10-10.5
Density, lbs/ft^3	2 8
Effect of freezing	None
Shelf life	1 year

All recommendations for metering dilute paste size emulsion and using it with alum in paper sizing, as outlined previously, also apply to the use of dry size.

Compatibility of Saponified Rosin Size

Saponified rosin size of all types, paste, fluid, or dry, is strongly anionic. It is precipitated by hard water, alum, bivalent, and trivalent salts. It is also precipitated when mixed with cationic additives. Care should be taken to keep the addition points of rosin size, alum, and cationic additives well-separated. The rosin size should be reacted with alum in the stock before any other cationic additives are added.

Spills of Saponified Rosin Size

Spills of paste rosin size are insoluble in cold water and slowly soluble in hot water. The preferred procedure for handling spills is to soak up in sawdust or other absorbent material and remove to a land fill. All soaps are highly toxic to fish and care should be taken that runoff from the landfill area does not contaminate streams. If this is a potential problem, the spill should be dissolved with hot water and steam lances and run to the sewer for treatment in a waste treatment system.

Spills of fluid and dry rosin size may be hosed to the sewer with cold or preferably warm water. All saponified sizes are biodegradable and will not affect waste treatment system operation. After suitable biological treatment, the effluent is no longer toxic to fish.

Safety Considerations with Saponified Rosin Size

There are no significant safety hazards associated with the handling of saponified rosin size. However, prolonged contact of saponified rosin sizes with the skin in concentrated or dilute form can result in minor skin irritation. Rubber gloves are recommended.

Hot paste size should be handled with the same precautions as any hot liquid. Protective goggles and gloves are recommended.

Improper handling of dry rosin size can produce a dust which is irritating to the eyes and throat. Handling methods should minimize dusting. Proper ventilation of the work area is recommended.

Government Regulation of Saponified Rosin Size

Manufacturers of saponified rosin size should certify that ingredients are on the EPA TSCA inventory.

Saponified rosin size is considered to be non-hazardous under OSHA regulations. However, as a matter of good plant practice, an MSDS should be obtained from the supplier.

Saponified rosin size claimed by the manufacturer to meet FDA requirements is suitable for use in food packaging paper and paperboard.

Dispersed Rosin Size

Dispersed rosin size is prepared by the emulsification of rosin in water. Dispersed size contains essentially 100 percent unsaponified rosin. Typical properties of dispersed rosin size are shown in Table 4. Dispersed rosin size is more efficient than saponified rosin size. Thus, a dispersed rosin size may replace saponified rosin size entirely for cost reasons. However, there are important secondary advantages in the use of dispersed rosin size in regard to improved paper strength, lower alum requirements, less pH sensitivity, and ability to size paper up to approximately pH 6.5 under proper conditions. In recent years, dispersed rosin size has replaced saponified rosin size in many mills.

Storage and Handling of Dispersed Rosin Size

Dispersed rosin size is shipped by the manufacturer in tank trucks, tank cars, and drums. The product must be protected from freezing during transit. Dispersed sizes are stored at room temperature and heating is never recommended.

Table 4. Typical Properties of Dispersed Rosin Size

Total solids, weight %	3 5
Appearance	White, fluid emulsion
Viscosity at 77°F, cps	5-25
Density at 77°F, lbs/gal	8.7
pH	5.5-6.5
Effect of freezing	Breaks
Shelf life	1 year. Some solids may settle out during prolonged storage.

All emulsions of solid, water-insoluble materials can be broken by excessive shear, which causes the particles to collide and coalesce. For unloading dispersed sizes, only low-shear equipment should be used. Air padding, or low-speed, low-shear centrifugal pumps, or diaphragm-type pumps are recommended. Gear or rotary type pumps are not suitable. Manufacturers of dispersed size can make recommendations for unloading and handling pumps for their products.

Due to its acidic pH, only stainless steel, tile, plastic, or plastic-lined tanks and equipment are recommended for handling dispersed size. Type 304 stainless steel is preferred because of the necessity of regularly cleaning the equipment with hot alkaline solutions.

Metering Dispersed Rosin Size

Dispersed rosin size can either be diluted to 2-5 percent solids for metering, or it can be metered in concentrated form and then diluted. Water used for diluting dispersed size must be below 50 ppm hardness and 50 ppm alkalinity to produce a stable dilute dispersion. Softened water will be required in most locations.

Dilution systems for dispersed rosin size may be batch systems operated manually, or automatic dilution units available from some size manufacturers. Dilute dispersed size requires the same materials of construction as concentrated dispersed size. Low-shear transfer and metering pumps, recommended by the dispersed size manufacturer, must be used for handling dilute dispersed size.

Metering devices consist of magnetic flowmeters, orifice-type differential pressure meters, rotometers, and others. Volumetric diaphragm pumps with variable-speed controls are also used.

Using Dispersed Rosin Size in Sizing Paper

The anionic dispersed rosin size particles are precipitated by alum and become cationic for retention on the fiber. Further reaction with alum occurs in the dryer section to produce a hydrophobic mixture of free rosin and aluminum resinate on the fiber surface.

Dispersed rosin size is usually added at the same point as alum (simultaneous addition) or after the point of alum addition (reversed addition). A minimum alum level of 0.75 percent is recommended and the headbox usually has a pH 4.7-5.2. By use of suitable cationic retention aids, dispersed rosin sizes can be used at headbox pH up to 6.5 and at alum levels as low as 0.5 percent. Under these conditions, strength and operability advantages approaching those of alkaline sizing materials are obtained.

Major dispersed rosin size manufacturers provide technical service to optimize performance of their products in different grades and under varying paper mill conditions.

Compatibility of Dispersed Rosin Size

Dispersed rosin size is highly anionic. All previous comments concerning saponified size compatibility apply to dispersed size.

Spills of Dispersed Rosin Size

Spills of dispersed rosin size are water-dilutable and can be hosed to the sewer. The material is biodegradable and will not cause any problems in waste treatment systems.

Safety Considerations with Dispersed Rosin Size

Dispersed rosin size does not present a safety hazard in normal use. Due to its slightly acidic pH, care should be taken to avoid contact with the eyes and on the skin.

Where danger of splashing in the eyes or on the skin exists, rubber gloves and protective goggles are recommended.

Government Regulations of Dispersed Rosin Size

Manufacturers of dispersed rosin size should certify that all ingredients are listed in the EPA TSCA inventory. Dispersed rosin size is usually classified as non-hazardous under OSHA regulations. However, as a matter of good plant practice, an MSDS should be obtained from the supplier. Dispersed rosin size claimed by the manufacturer to meet FDA requirements is suitable for use in food packaging paper and paperboard.

Reactive Sizing Agents

Types of Reactive Sizing Agents

Reactive sizing agents, also called alkaline sizes, have been given this designation because they chemically react with cellulose, typically in the pH 6-9 region. Reactive sizes can be used with alkaline fillers, such as calcium carbonate. Alum is not required for the sizing reaction. All reactive sizes contain a highly reactive group, which is capable of reacting with cellulose, and also a hydrophobic group which produces the necessary water repellency.

A cellulose reactive size is not naturally cationic and is retained on the fiber by use of cationic retention agents. Actual reaction with the cellulose takes place during the drying operation. The sizing agent spreads on the fiber and orients itself with the reactive group in contact with the cellulose and the hydrophobic group oriented outward. Final reaction with the cellulose, in the nearly-dried sheet, produces a covalent bond of the hydrophobic group to the cellulose molecule.

The use of cellulose-reactive sizes allows higher levels of filler, use of calcium carbonate as filler, increased use of secondary fiber, and more extensive reuse of white water. Additional advantages gained through the use of reactive sizes vs. rosin size include the following: higher levels of sizing, increased dry strength, improved tear strength, reduced energy usage, reduced corrosion, reduced fresh water requirements, easier effluent treatment, and increased production.

Reactive sizing agents are used in writing and printing papers, unbleached kraft grades, and bleached kraft specialties. They are also used in food packaging applications requiring prolonged resistance to strong penetrants. An important, although minor use, is in permanent papers requiring high-extract pH to resist deterioration of physical properties on long-term storage.

Alkylketene Dimer Size

Alkylketene dimer (AKD) size is usually sold by the manufacturer in prepared emulsion form. The emulsions range from essentially uncharged emulsions, requiring additional retention aids, to strongly cationic emulsions which are self-retaining. Typical properties of AKD emulsions are shown in Table 5.

Shipment and Storage of Alkylketene Dimer (AKD) Emulsion

AKD emulsion is shipped in tank trucks or in standard fiber or steel drums. Tank car shipments are rarely made because the perishable nature of the product requires use within the 30-day shelf life.

Due to the acidic pH of 3.0-4.5, AKD emulsion must be handled in stainless steel, acid-resistant tile, plastic, or plastic-lined tanks, piping, and pumps. Type 316 stainless steel is preferred. Tanks should be located indoors, if possible, to prevent freezing in

Table 5. Typical Properties of Alkylketene Dimer (AKD) Emulsions

	Uncharged	Cationic
Total solids, weight %	6-12	15
Appearance	White, fluid emulsions	White, fluid emulsions
pH	3.0-4.0	3.0-4.5
Viscosity at 77°F, cps	10-50	10-50
Density at 77°F, lbs/gal	8.5	8.6
Effect of freezing	Breaks	Breaks
Shelf life at 90°F	30 days	30 days
Zeta potential of emulsion, millivolts	+3 to -3	+23 to +28

cold climates, or excessive heating of contents in warm climates. Refrigeration may be required in extremely hot areas to keep the product temperature in storage at the preferred 75°F (24°C) or below. Unloading lines should extend to the bottom of the tank to prevent foaming. The tank should be covered to protect from contamination and to minimize surface evaporation.

Pumping and Metering AKD Emulsions

AKD emulsion is not shear-sensitive and may be handled in most types of centrifugal, rotary, gear, or diaphragm pumps. Open impeller centrifugal pumps are preferred for transfer. Diaphragm pumps are recommended for metering. The emulsion manufacturer should be contacted for specific recommendations.

AKD emulsion is normally handled and metered in concentrated form. After metering, the emulsion should be randomly diluted with water 10:1 before adding to the stock. Emulsion flow may be controlled by means of a rotometer and a manual control valve, although a magnetic flowmeter, automatic control valve, and recorder is preferred. A

20-40 mesh final filter or strainer is recommended at the point of addition.

Using AKD Emulsion in Sizing Paper

AKD emulsion is always added close to the paper machine, with actual points of addition ranging from the accept side of the stuff box to the headbox line ahead of the stream flow valve. Retention aid, when used, is usually added after the screens. Alum, when used, is usually added at the fan pump.

The rate of sizing development of AKD emulsion is dependent upon the hot extract pH of the paper (TAPPI Method T435). Minimum extract pH is about 6.5 and pH 7.0-8.5 is preferred. This usually requires a headbox pH of 6.3 or higher. The reaction of AKD with the cellulose begins in the dryers ahead of the size press. Sufficient sizing is always developed for normal size press operation. However, the application of starch at the size press may cover the partially reacted ketene dimer groups and result in less sizing at the reel. In order to estimate the final level of sizing with AKD emulsion, the reel sample of paper or board is given a moderate oven cure, usually 5-10 minutes at 220°F

(104°C), before testing for sizing. Alternatively, paper may be tested off the winder, without an oven cure.

Since AKD size is capable of good distribution over the fiber surfaces, and reacts with cellulose on a molecular basis, sizing efficiency is extremely high (see Figure 1). Usage levels range from 0.05 percent in slack-sized grades to about 0.5 percent in hard-sized grades.

Compatibility of AKD Emulsion

The uncharged AKD emulsion may be incompatible when mixed with either strongly cationic or strongly anionic additives. Cationic AKD emulsion is compatible with most cationic additives, but incompatible when mixed with anionic additives. Good manufacturing practice suggests that all strongly charged additives be added through separate lines, physically separated from each other by several feet at the addition point to avoid the danger of accidental mixing before they are thoroughly blended with the stock.

AKD Emulsion Spills

AKD emulsion spills are readily dilutable with cold water and should be hosed to the sewer. The products are biodegradable and will not cause problems in waste treatment systems.

Safety Considerations with AKD Size

Uncharged AKD emulsion does not present a health hazard in normal use. Cationic AKD emulsion contains low levels of residual epichlorohydrin and label warning instructions should be followed.

Due to an acidic pH, all AKD emulsions may be slightly irritating to the skin and eyes. Use of rubber gloves and protective goggles is recommended.

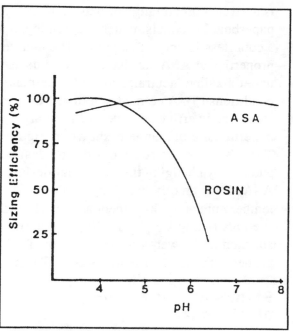

Figure 1. Effect of pH on Sizing Efficiency

Government Regulations of AKD Size

The manufacturer of commercial AKD emulsion should certify that all ingredients are on the EPA TSCA inventory.

Uncharged AKD emulsion is considered nonhazardous under OSHA regulations, while cationic AKD emulsion is considered hazardous. An MSDS on the particular product being used should be obtained from the manufacturer.

AKD emulsion claimed by the manufacturer to meet FDA requirements is suitable for use in the manufacture of food packaging paper and paperboard.

Alkenyl Succinic Anhydride (ASA) Size

Alkenyl succinic anhydride (ASA) size[1]

---

[1]Information on alkenyl succinic anhydride size was contributed by Robert T. McQueary, National Starch and Chemical Corporation, and Walter F. Reynolds, American Cyanamid Corporation.

is used in a broad range of paper and paperboard products, which are employed in a countless variety of end uses. The unique properties of ASA size have allowed its use under alkaline, neutral, and acid papermaking conditions (5). A comparison of an ASA and rosin size in efficiency over the normal papermaking pH range is shown in Figure 1. The ASA size can be designed for a desired end use by changing the molecular design. ASA size is an oily liquid at room temperature and 100 percent active. The ASA is emulsified in the paper mill just prior to addition to the wet-end stock system. Typical properties of ASA size are listed in Table 6.

Shipment and Storage of ASA Size

ASA size is shipped in drums, polyethylene-metal semi-bulk containers, tank truck, or tank car. With proper storage, the ASA is noncorrosive, stable, and unaffected by typical storage temperatures. Even though it may increase in viscosity at temperatures below 50°F (10°C), it can be used upon returning to a normal temperature range. The ASA can be stored in a covered plain steel tank and is stable, provided it is protected from moisture. In a bulk tank, this is readily achieved by using a dry air pad.

ASA is very active and can react with many materials. Should it react with water, a corresponding acid may be formed which could be mildly corrosive. However, the small quantity that may be encountered can be utilized in a properly designed wet end system to obtain sizing. This has never been a problem, even in bulk systems, which have been in use since the late 1960s. ASA products are oily and do not freeze. They have a pour point ranging from 35 to 45°F (2-7°C).

Emulsification of ASA Size

Historically, most sizes have been emulsified in the mill prior to their addition to the papermaking process. ASA size must also be emulsified. It can be furnished in either the nonemulsifiable (chemical emulsifier needed) or the emulsifiable (by mechanical means) form. Emulsification allows the size particles to be readily mixed with the fiber slurry and evenly distributed upon the fibers to optimize sizing and retention.

The ASA may be emulsified by several methods. One method incorporates a high-shear turbine pump to achieve emulsification of the ASA size (6). Other emulsifiers using high-pressure, high-shear, or both, are high-speed agitators and mechanical or ultrasonic homogenizers.

Another method employs a chemical activator which is thoroughly mixed with the ASA prior to emulsification (7). The activator enables a fine, consistent particle size emulsion to be formed in low-pressure and low-shear conditions. Low-shear forces, adequate to emulsify the ASA using this new technology, are produced by moderate-speed agitation or passing the mixture through a mixing valve, common aspirator, or mechanical orifice. The equipment available for emulsifying the ASA is extremely simple, provides excellent control, and requires little maintenance during essentially continuous operation. The emulsifiers are designed to run manually, fully automated, or even computerized. Emulsification units can be purchased from the ASA suppliers.

Either water or cationic starch solution can be used as the emulsification medium for ASA size. Cationic starch continues to be the most widely used, most efficient, and most effective. Cationic starch provides emulsion stability and uniformity, contributes significantly to the overall retention, and enhances the physical properties of the paper

Table 6. Typical Properties of Alkenyl Succinic Anhydride (ASA)	
Total solids, weight %	100
Appearance	Clear, amber oil
Viscosity at 77°F, cps (Brookfield at 20 RPM)	50-500
Density at 77°F, lbs/gal	7.8-8.0
Effect of freezing	Freeze/thaw stable
Shelf life	Indefinite, if protected from moisture.

Note: ASA are oily products which do not freeze. They have a pour point ranging from 35 to 45°F.

or paperboard in the areas of printability, internal strength, and dry strength.

Either a pregelatinized, cationic starch soluble in water at room temperature, or a conventional cationic starch, requiring high-temperature cooking, is suitable. If hot cationic starch is used, it should be cooled to avoid excessive hydrolysis of the ASA during emulsification. The cationic starch solutions should be protected from bacteriological attack.

The use of a cationic donor in the emulsification medium has been proven beneficial to the ASA size efficiency. The ASA size is generally emulsified with one-half to three parts of cationic material to one part ASA (8). Among the cationic materials that may be employed for this purpose are long-chain fatty amines, amine-containing natural or synthetic polymers, substituted polyacrylamide, animal glue, cationic thermosetting resins, and polyamide-epichlorohydrin polymers (9).

Each emulsification system should be designed to maintain high-performance without hydrolysis. The stability and the related efficiency of the emulsion is a function of many variables. A properly

designed emulsification system can produce an emulsion that is stable for several hours.

The emulsion can be made continuously or batchwise, depending on the specific mill location. If the emulsion is made continuously, a back-up emulsifier is needed to ensure uninterrupted supply to the paper machine. If a batch system is used, a hold tank capable of supplying the paper machine during normal maintenance is required.

Once a mechanical emulsion has been made, it should be added to the stock as soon as possible, preferably by direct addition, without going through a storage tank. Where this cannot be accomplished, the emulsion temperature should be brought below 104°F (40°C) using cool dilution water or starch solution, and the emulsion should be kept at pH 3.5-4.5 to minimize hydrolysis.

Pumping and Metering ASA Size Emulsions

The supply and metering system components depend on whether the emulsion is fed directly from the emulsifying unit to the paper machine or held in a storage tank. Feeding systems for direct addition do not require additional metering devices since the feed rate of the ASA size is set by the constant displacement pump supplying the ASA size to the emulsifier. Post-dilution of the emulsion has no effect on the feed rate, as long as the emulsion goes directly to the paper machine.

Stainless steel or plastic equipment is used for the storage and handling of ASA emulsion. If a pump is required for the addition of the emulsion to the machine, then centrifugal, rotary, gear, or diaphragm pumps can be used. A constant displacement pump, a magnetic flowmeter or a rotometer can be used as a metering device.

For more detailed information on preparing and handling the emulsion, ASA supplier assistance should be requested.

Using ASA Size in Sizing Paper

ASA emulsion is typically diluted prior to the addition to the fiber slurry to ensure proper distribution. Usual points of addition are: the inlet to the fan pump, stock chests, outlet of the refiners, or the inlet of the centriscreens (low-pressure screens). Some ASA systems have been utilizing thick stock emulsion addition with great success for over 5 years.

ASA size is employed at an addition rate of 0.05-0.5 percent of the dry weight of pulp, depending upon the furnish and the paper's end use requirements. Typically, ASA usage is 0.1-0.15 percent (2-3 lb/ton).

Some systems require ASA emulsion to be added to the stock close to the paper machine for two reasons:

1. To avoid prolonged exposure of ASA to the water phase of the stock,

2. To avoid the adverse effect of high-shear equipment.

In these systems, the inlet side of low-pressure screens (centriscreens) provides a short, but adequate, mixing and contact time

for good retention without encountering the high-shear force of centrifugal cleaners. Where these cleaners are not used, the inlet side of the fan pump provides a good point of addition.

With any size material, especially ASA, it is essential that the first pass retention be as high as possible to maximize the efficiency of the size and the paper machine operation. The use of a high-efficiency retention aid is recommended to attain maximum first pass retention. The choice of retention aid, anionic or cationic, will depend on local mill conditions and can best be determined by on-machine evaluation. The use of alum is permissible and, in many instances, is helpful in attaining good runnability. Some factors minimizing press roll picking to acceptable levels include high first-pass retention, use of alum, and maintaining a film of water on the press roll surface.

The relatively high degree of reactivity of the ASA size enables it to develop sizing rapidly. ASA is liquid at room temperature and in a state capable of reacting with cellulose. The size only requires proper distribution to impart sizing to the sheet. The emulsion readily opens up with a minimum of energy to flow and spread across the fiber to develop sizing.

To the papermaker, controlling the moisture pickup at the size press, coater, or subsequent rewetting operation is essential for optimum operation of the paper machine or auxiliary operations. With ASA size, pre-size press sizing is obtained in all grades using the typical amount of energy in the main drier section. ASA makes it possible to raise the pre-size press moisture without affecting the sheet moisture or increasing the energy required in the after-section driers. The uniform sizing enhances overall paper quality and production. Accelerated cure of the sheet is not needed to predict the ultimate sizing.

ASA provides efficient sizing for many applications. ASA, as a class of sizing agents, is capable of producing extremely hard sized paper or board. Each system can be designed to meet the end use and test requirements. It is also possible to use the ASA in surface treatments with the proper system design. An example of surface treatment is surface sizing at the size press or on the calender (10).

Compatibility of ASA Size

ASA size is compatible with common papermaking additives such as wet-strength resins, slimicides, dispersants, starches, natural gums, polyvinylamine retention aids, most defoamers and the like, when these are used at normal addition levels. The ASA suppliers are prepared to evaluate the compatibility of various materials used and make recommendations regarding the suitability of the materials.

Spills of ASA Size

In accordance with good housekeeping practice, spills of ASA size should be cleaned up as soon as possible. Large spills can be absorbed with absorbent clay or other suitable material commonly used for oil spills. The waste should be disposed of in accordance with federal and state regulations covering solid waste disposal.

Automatic emulsification systems have been designed to minimize the amount of effluent. Operation in both acid and alkaline paper mills utilizing ASA have shown that materials in the effluent can be handled normally. The starch emulsion materials are readily biodegradable. Disposal of waste emulsion can be handled by normal effluent treatment. Direct discharge of the emulsion should be avoided.

Waste ASA products which have not been contaminated with other substances are not hazardous wastes according to current Resource and Conservation Recovery Act (RCRA) guidelines.

Safety Considerations with ASA Size

ASA has been safely used in the manufacture of paper and paperboard for over 15 years.

ASA is a slight to moderate skin and eye irritant. Repeated or prolonged skin contact should be avoided. Gloves and goggles are recommended and normal precautions associated with the handling of weak acids should be followed. As with any acid, ASA should not be mixed with strong oxidizing agents. The chemical activator is an eye irritant and contact with the eyes should be avoided. Normal precautions should be followed and goggles should be worn.

No precautions are needed for handling the emulsion, except for eye protection. Flushing with water should be adequate for exposure to ASA or the emulsion.

Government Regulations of ASA Size

The manufacturer of the ASA size should certify that all ingredients are on the EPA TSCA inventory. The MSDS on the size should be obtained from the manufacturer. Suppliers involved in or contributing to the manufacture of food packaging paper and paperboard should certify that products are in compliance with FDA requirements.

Wax Emulsion Sizing Agents
Types of Wax Emulsion Sizing Agents

Waxes are highly hydrophobic materials which, when applied in emulsion form, are suitable for use in the sizing of paper and paperboard. Paraffin wax is most commonly used in wax emulsions for paper sizing, although specialized waxes such as microcrystalline or synthetic waxes may occasionally be used.

Wax emulsions are usually classified according to their chemical stability as being either acid-stable or acid-breaking. Acid-stable emulsions are stable in the presence of moderate levels of acids, alkalies, alum, and other bivalent and trivalent metallic salts. Acid-breaking emulsions are unstable to acid, alum, or other bivalent or trivalent metallic salts. Most wax emulsions used in paper sizing applications are of the acid-stable type.

Properties of Wax Emulsion

Typical properties of acid-stable and acid-breaking wax emulsions are seen in Table 7.

Storage and Handling of Wax Emulsion

Wax emulsion is shipped in tank trucks, tank cars, and drums. Because of the high solids and hard waxes used, wax emulsion is extremely shear-sensitive. Unloading of tank trucks and tank cars is preferably carried out by air padding. Unloading pumps for wax

emulsion must be low-speed, open-impeller centrifugal pumps or diaphragm pumps having large check valve clearances. Other pumps are not suitable. Consult the wax emulsion manufacturer for specific pump recommendations for wax emulsion unloading.

Due to the acidic pH of most wax emulsions, bulk storage tanks should be constructed of fiberglass-reinforced plastic, stainless steel, or tile. Unlined steel tanks should never be used for bulk storage of wax emulsion at paper mills because of problems with wax breakout and emulsion discoloration.

Bulk storage tanks for wax emulsion should be located indoors and as close as possible to the point of use. Outside storage tanks for wax emulsion are not recommended because of the necessity of keeping the emulsion at about 70°F (21°C) for good pumping stability, and to avoiding freezing under winter conditions.

When wax emulsion in bulk tanks must be located outside, the tank pumps and piping should be equipped for hot water heating and heavily insulated. Do not use steam for

Table 7. Typical Properties of Wax Emulsions

	Acid-Stable	Acid-Breaking
Total solids, weight %	40-55	50-60
Appearance	White, fluid emulsions	White, fluid emulsions
pH	6-5	9-10
Viscosity at 77°F, cps	300-1000	300-1000
Density at 77°F, lbs/gal	8.0	7.6
Effect of freezing	Breaks	Breaks
Shelf life	6 months. May separate on prolonged storage.	6 months. May separate onprolonged storage.

heating or tracing because localized overheating may break the emulsion.

The buildup of wax on the surface of wax emulsion in storage can be greatly reduced by storing at high relative humidity. This can be accomplished in bulk storage tanks by humidification of air space over the liquid. Commercial humidification systems are available.

Piping for handling wax emulsion should be plastic, plastic-lined, or stainless steel. Plastic piping should be able to withstand a temperature of 175°F (79°C) for hot water flushing of the system to remove deposits.

Piping for bulk wax emulsion systems should always be oversized to minimize pressure and shear on the emulsion. Use three-inch pipe for bulk unloading. Piping for handling wax emulsion in a metering system should be at least one inch, preferably 1 1/2-inch to ensure minimum pressure drop through the system.

Metering Wax Emulsion

Wax emulsion may be metered to the paper machine in concentrated form, or it may be diluted before use. When concentrated wax emulsion is metered, the distance between the bulk storage tank and the point of use should be as short as possible. Suitable metering pumps are plunger-type pumps with large clearances between the plunger and the cylinder, and diaphragm-type pumps. Both plunger-type and diaphragm-type pumps should be equipped with check valves having large clearances to avoid placing excessive shear on the emulsion. Flows may be regulated by manual or automatic adjustment of stroke length, or by manual or automatic adjustment of stroking speed.

Where the distance between the wax emulsion bulk storage tank and the point of use is greater than 50-75 feet, or where multiple points of addition are required, a dilution system should be used. Wax emulsion becomes more stable to pumping and handling as the solids are reduced. Wax emulsion manufacturers can recommend specific designs for batch dilution systems, or continuous automatic dilution systems. Softened water is usually required for diluting acid-breaking emulsions.

Due to the high-solids content, wax emulsion is prone to the formation of a wax film on the surface during storage and handling. When dried, this film forms small wax flakes floating on the product. Since it is virtually impossible to keep small amounts of wax breakout from forming, wax emulsions should always be filtered with a 20-40 mesh filter at the point of addition to the paper stock.

Concentrated wax emulsion should be diluted with a stream of water ahead of the filter. Since wax particles can also affect the efficiency of check valves on plunger and diaphragm-type pumps, it is recommended that a basket-type strainer be installed ahead of metering pumps. The strainer should be 20 mesh, stainless steel, and large in area to avoid placing excessive shear on the emulsion.

Using Wax Emulsion in Sizing Paper

Wax emulsion is usually used as a supplement to rosin size, primarily in hard-sized grades. In recent years these hard-sized applications have been replaced in many cases with cellulose-reactive size. Internal use of wax emulsion is usually in the range of 0.1-0.3 percent, dry basis. The wax emulsion is added close to the paper machine, preferably after the rosin size and alum addition. The most common point of addition is the fan pump intake.

Acid-stable wax emulsion is most commonly used in paper grades. This anionic emulsion is retained by cationic aluminum resinate, cationic alumina polymer, or by the use of synthetic cationic retention aids.

Acid-breaking wax emulsion is usually used only where there is insufficient alum or other cationic material present to ensure the retention of an acid-stable emulsion. Acid-breaking emulsion may be added to the thick stock before alum addition or at the fan pump after alum addition. Because of the sensitivity to precipitation by alum and hard water salts, care must be taken in the use of acid-breaking emulsion to prevent premature coagulation and formation of wax spots in the sheet. It is for this reason that the more reliable acid-stable wax emulsion is usually used in paper.

Wax Emulsion Compatibility

Both acid-stable and acid-breaking wax emulsions are anionic in character and will be precipitated when mixed with cationic additives. The acid-stable emulsion has good stability to hard water, alum, and acids, while the acid-breaking emulsion is highly-sensitive to these materials. The addition points of both acid-stable and acid-breaking emulsions should be physically separated from any cationic additives.

Wax Emulsion Spills

Wax emulsion spills are readily dilutable with cold water and can be hosed to the sewer. The products are biodegradable and will not cause problems in waste treatment systems.

Safety Considerations with Wax Emulsion Size

Wax emulsion does not present a health hazard under normal conditions of use. These mildly acidic or alkaline emulsions may be irritating if splashed in the eyes and protective goggles are recommended.

Government Regulations of Wax Emulsion Size

The wax emulsion manufacturer should certify that wax emulsion ingredients are on the EPA TSCA inventory. Wax emulsion is classified as nonhazardous under OSHA. However, good plant safety practice suggests that an MSDS be obtained from the wax emulsion manufacturer for the specific product being used. Wax emulsion size claimed by the manufacturer to meet FDA requirements is suitable for use in food packaging grades of paper and paperboard.

Literature Citations

1. "Internal Sizing of Paper and Paperboard," *TAPPI Monograph Series No. 33*, Atlanta, GA: TAPPI PRESS, 1971, pp. 170-187.

2. "Internal Sizing of Paper and Paperboard," *TAPPI Monograph Series No. 33*, Atlanta, GA: TAPPI PRESS, 1971, Chap. 4, pp. 54-93.

3. *Sizing of Paper*, 2nd edn., Atlanta, GA: TAPPI PRESS, 1989.

4. Casey, James P., *Pulp and Paper Chemistry and Chemical Technology*, 3rd edn., New York: Wiley-Interscience, Vol. 3, Chap. 16, pp. 1547-1588.

5. Mazzarella, E. D., L. J. Wood, and W. Maliczyszyn, U. S. Patent 4,040,900 (August 9, 1977).

6. Wurzburg, O. B., and E. D. Mazzarella, U. S. Patent 3,102,064 (August 27, 1963).

7. Mazzarella, E.D., L. J. Wood, and W. Maliczyszyn, U. S. Patent 4,040,900 (August 9, 1977).

8 Wozniak, A. M., *1982 Papermaker's Conference Preprint*, p. 53.

9. Mazzarella, E. D., L. J. Wood, and W. Maliczyszyn, U. S. Patent 4,040,900 (August 9, 1977).

10. Kennedy, A. E., U. S. Patent 4,311,767 (January 19, 1982).

Chapter 6

Surface Sizing Agents

by Richard D. Harvey
and
T. Small

Introduction

The purpose of surface sizing is multifaceted with the primary objective being to cement fiber to the body of the paper; other objectives include decreasing porosity, increasing oil resistance, and increasing paper strength qualities.

Several additives are used in this application including starch, latex, polyvinyl alcohol (PVA), carboxymethyl cellulose (CMC), and animal protein. Starches represent over 90 percent of the additives used in this application and, therefore, will be the concern of this chapter.

General Description
General Composition

Starch is a polymer with glucose as the basic repeating unit. Two types of polymers are present: linear or straight chain (amylose), and branched chain (amylopectin). The molecular weight (size) and ratio of amylose-amylopectin molecules in starch will vary as a result of the following factors:

1. Source of starch such as corn, potato, heat, tapioca.

2. Genetic variety as in corn, regular yellow dent, waxy maize, or high amylose.

3. Type of modification.

Starch is supplied to the papermaker in a dry form. Commercial corn starch contains 10 percent moisture. The equilibrium moisture content varies with the source generally determining the standards for different products. Commercial potato starch, for example, contains 15 percent moisture. The particle size of the dry starch varies relative to processing methods and handling and ranges from fine powdered (75 percent minus 200 mesh) to large particles, often called pearl starch.

Contained within each particle are discrete granules which have characteristics of size, shape, light transmission, and adsorption properties relative to their source. Granular starch in an aqueous system is referred to as a suspension or a slurry.

Starch in the granular form is of little or no value for surface sizing and would function primarily as a filler. It is necessary that starch be cooked to obtain the functional qualities desired. The cooking process is referred to as gelatinization, pasting, or hydration.

Starch Modifications

Many different commercial starches are available including unmodified, buffered, acid-modified, oxidized, and derivatized. Starch modifications are conducted for many purposes such as pH control, improved gelatinization properties, viscosity control, improved filming, reduced retrogradation, which causes viscosity increase. Modified industrial starches are covered in 21 Code Federal Regulations, paragraph 178.3520.

Common starch modifications include:

1. **Buffering.** Buffering consists of simply incorporating additives in the starch. The most common additives are for pH control to enhance enzyme activity and to optimize storage properties. Starches are also treated with additives to buffer against oxidation, color formation, and foaming.

2. **Acid Modification.** Starch in a slurried form is treated with acid, generally sulfuric or hydrochloric, to reduce the starch molecular weight and viscosity.

3. **Oxidation.** Starch in a slurried form is treated with an oxidizing agent, generally sodium hypochlorite. The oxidation reaction has several effects on the product including: hydrolysis or reduction of molecular carboxyl groups. The carboxyl group starch. This reaction is generally conducted in a highly alkaline system. However, variations in alkalinity result in changes in the product properties.

4. **Derivatization.** Starch in a slurried form is treated with chemical reagents resulting in pendant groups on the starch molecule. The chemical bond between the pendant group may be either an ester or an ether with the following formulas:

$$Ester \quad (starch-O-\overset{\displaystyle O}{\overset{\displaystyle \|}{C}}-R)$$

$$Ether \quad (starch-O-R)$$

The ether linkage is generally preferred because it is much more durable. The most common derivatized starch is ethylated or hydroxyethyl ether in which R is:

$$\begin{array}{ccc} H & & OH \\ | & & | \\ C & \!\!-\!\!-\!\!- & C-H \\ | & & | \\ H & & H \end{array}$$

Derivatization generally improves cooking properties and storage qualities. Selection of a specific product will be influenced by many factors including volume used, type of equipment available, condition of equipment available, as well as specific objectives and personal preferences.

Storage and Handling
Shipping Containers

Starch is available in bags, bulk truck, or bulk rail cars. Bulk storage should be as close as is practical to the unloading point and the preparation area. Starch, being dry, is tolerant to large variations in temperature and can be stored at ambient conditions including subzero temperatures.

Bulk Storage and Handling

Corn starch is a dry, powdery material which can be stored and handled automatically with excellent results. The most common bulk delivery vehicle is the

Airslide® [1] rail car. There are other system designs which will be covered later. Starch bulk handling systems for Airslide® rail cars should be designed for conveying material up to 40 lbs/ft^3, which requires a velocity of 6,000 fpm.

Figure 1 illustrates basic features of a bulk handling and storage system. Numbers in brackets [ ] refer to points in Figure 1.

Bulk starch is received in an Airslide® rail car [1] and unloaded through a vacuum unloading sled [2] into the conveying line which leads to the vacuum filter [3]. The unloading blower package [4] also supplies a positive pressure which transfers the starch from the vacuum filter [3] to the silo [5] and the conveying air exits through the silo filter [6]. The unloading blower package also produces low-pressure air for the Airslide® pads on the rail car to fluidize the starch. The starch is transferred from the silo [5] by positive pressure provided by the transfer blower package [8].The starch is metered into the conveying air with the rotary valve [7].

The slurry makedown system consists of a continuously agitated slurry makedown tank [12], a metered water supply [13], and a weigh hopper [10]. The slurry tank is first filled with the correct amount of water [13], and starch is conveyed into the weigh hopper [10]. When the weight is reached, the silo rotary valve [7] stops and the blower [8] runs until the line is empty and then shuts down. The rotary valve [11] then transfers the starch into the slurry tank [12]. When all the starch has been transferred and wetted out, the batch is pumped to another tank and another batch can begin.

---

[1]Airslide ® is a registered trademark of Fuller Company, Catosauqua, Pennsylvania.

Other systems would use a pressure-conveying sled under the Airslide® rail car in place of the vacuum sled [2]. This system would send the starch directly into the silo [5].The rail car could be a pressure differential (P/d) car where the car is pressurized, and the pressurizing air blows the starch into the silo. A bulk truck would connect to a separate line to the top of the silo. A silo instead of Airslide® pads could have a bin discharger which moves and helps the starch flow to the silo outlet.

Construction Materials

Steel construction is commonly used for the silo with a good protective exterior painted finish and a special interior coating of a phenolic resin or equal for a smooth rust-free surface. The conveying lines are usually aluminum tubing with long radius bends. The makedown equipment for the starch slurry system, and all other items contacted by the starch slurry must have stainless steel tanks and piping.

Items of Special Interest

1. Along with the proper electrical grounding, proper electrical classification for the controls and electric motors must be used for the wet or potentially dusty areas.

2. Sight glasses in the transfer and conveying lines indicate flow activity to the operator.

3. A knife gate or slide-type valve can be used above the rotary valves for ease of maintenance to the rotary valves.

4. Alternate starch slurry systems are available using weigh cells or load cells for weighing the water and the starch.

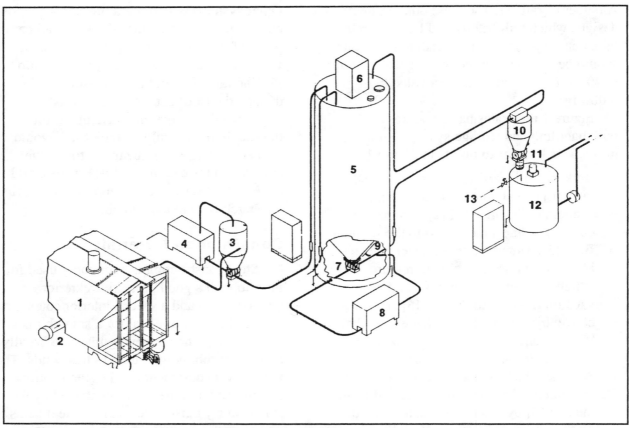

Figure 1. Bulk Handling and Slurry Preparation
(1) Airslide rail car, (2) Vacuum unloading sled, (3) Vacuum filter, (4) Unloading blower package, (5) Silo, (6) Silo filter, (7) Rotary valve (8) Transfer blower package, (9) Fluidizing air pad, (10)Weigh hopper, (11) Rotary valve, (12) Slurry tank, (13) Water supply

Paste Preparation

To be effective as a sizing agent, starch must be gelatinized and dispersed. The more complete the dispersion, the more effective the product, and the better the quality control.

Methods of paste preparation vary widely including simple batch cooking, batch enzyme conversion, continuous jet cooking, continuous enzyme conversion, and thermal-chemical conversion (TCC). The largest volume process is TCC followed by batch enzyme conversion.

The goal of preparing an effective starch sizing agent is generally monitored by three parameters: desired viscosity, desired concentration, and proper temperature.

In preparing starch paste, two factors significantly affect viscosity. First, dispersion is very important with more complete dispersion yielding lower viscosity. Second, starch molecular weight is important, with viscosity changing proportionately to changes in molecular weight.

Processing Premodified Starches
Batch and Jet Conversion

In the batch conversion process, slurried starch in a tank is heated by injection of live steam to a temperature range of 190-205°F (88-96°C) (atmospheric). The paste is held at

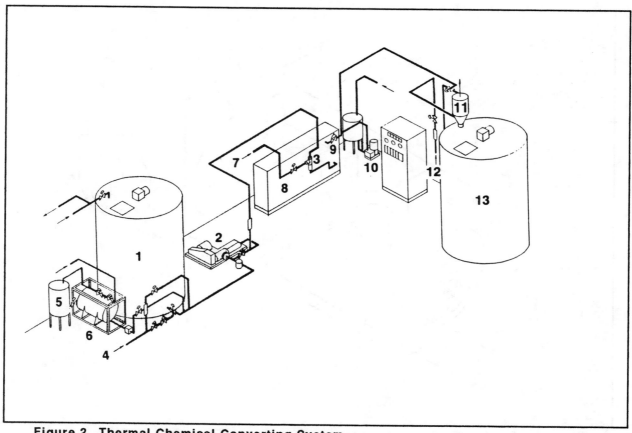

Figure 2. Thermal Chemical Converting System

(1) Slurry, (2) Displacement pump, (3) Jet, (4) Pre-dilution water, (5) Soluble oxidant, (6)Pressure tank, (7) Steam, (8) Coil, (9) Pressure regulating valve, (10) Alkali, (11) Flash chamber, (12) Post-dilution water, (13) Receiving tank or storage tank

the high temperature for a time to maximize dispersion.

In the jet conversion process, starch slurry is pumped through a jet into which live steam is injected resulting in instantaneous temperature rise to 230-325°F (110-163°C) Retention time at elevated temperature is negligible before flashing back to atmospheric condition. Several products are available which are premodified, including acid-modified and oxidized, to provide the desired molecular weight viscosity.

Processing Unmodified Starches
Thermal-Chemical Conversion

As previously indicated, the largest volume of starch for surface sizing is being prepared by thermal-chemical conversion (TCC). Two processes are conducted simultaneously in the thermal chemical conversion, gelatinization and dispersion; and hydrolysis or viscosity modification. The TCC process involves the following as detailed in Figure 2:

Slurry [1] is transferred by means of a positive displacement pump [2]

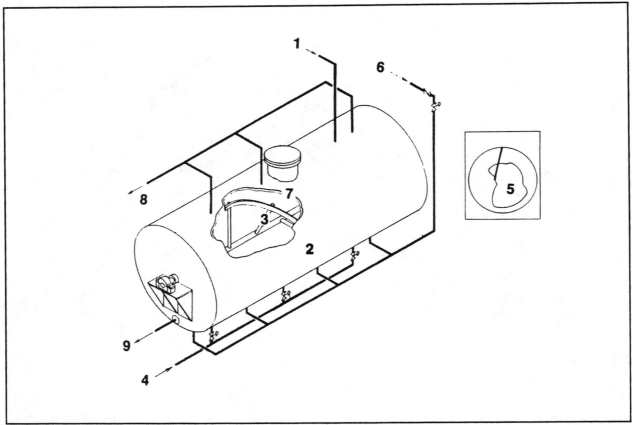

Figure 3.   Enzyme Conversion Process
(1) Starch slurry, (2)Conversion tank, (3) Agitator, (4) Cold water, (5) Temperature controller,
(6) Live steam injection, (7) Jacket, (8) Drain, (9) Discharge line

(progressive cavity) to a jet [3]. The slurry concentration can be regulated by controlled addition of pre-dilution water [4]. A soluble oxidant [5] (commonly ammonium persulfate) is added proportionately to the slurry by means of a pressure tank [6]. The oxidant, catalyzed by heat, hydrolyzes (reduces molecular weight of) the starch. At the jet, steam [7] is injected to instantaneously heat the slurry to the range of 305°F (152°C)pasting the starch. The starch paste is retained in the coil [8] for a period of approximately five minutes. Condensation and temperature are maintained in the coil by means of a pressure regulating valve [9] set at a pressure greater than the vaporization

point. Paste pH exiting the coil is regulated by the addition of caustic or other suitable alkali [10]. The steam is flashed off in the flash chamber [11] and the product reduced to atmospheric temperatures. Post-dilution water[12] is metered to the flash chamber to regulate solids and temperature. The product exiting the flash chamber is collected in the receiving tank or storage tank [13].

Construction Materials

All materials in contact with ammonium persulfate require stainless steel 300 series and it is the preferred construction material for the converter and starch piping: steam, air, water, and caustic supply.

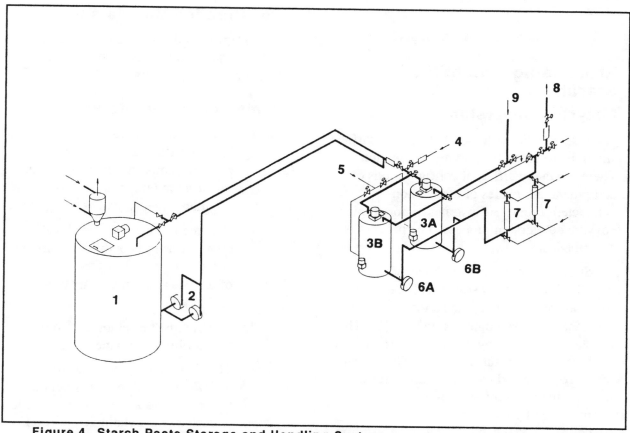

Figure 4. Starch Paste Storage and Handling System
(1) Storage tank, (2) Centrifugal pump, (3A) Run tank, (3B) Run tank, (4) Enroute water, (5) Live
steam injection, (6A) Centrifugal pumps, (6B) Centrifugal pumps, (7) In-line screening device,
(8) Size press, (9) Size press

Pumps for conveying starch must be positive-displacement types, preferably progressive cavity. Copper is acceptable for use only in air supply on pneumatic controls.

Items of Special Interest

1. Starch of choice for this process is normally unmodified, unbuffered (pH).

2. Variations in quality can affect hydrolysis.

3. Granular starch (slurry) settles rapidly without agitation and becomes dilatant.

4. Valves and piping should be designed as in Figure 1 to minimize potential for plugging and other handling problems.

5. Steam requirements are 150 pounds saturated steam properly trapped to avoid excessive dilution.

6. Pressure must be maintained in the coil to assure totally condensed conditions. Vaporization in the coil results in channeling, significantly reduced retention times, and potential carryover of soluble oxidant to the storage tank.

7. Proper pH control is essential to maintaining product quality.

Processing Unmodified Starch

Enzyme Conversion

A substantial volume of starch for surface sizing is prepared by enzyme conversion. There are many different configurations of equipment and cycles of conversion conducted. Figure 3 illustrates a more typical enzyme conversion process. The process is described as follows:

Starch slurry ranging in concentration of 20-40 percent dry solids with bacterial alpha amylase enzyme added, is transferred to a conversion tank [2]. The starch is mixed by an agitator [3] (ribbon, blade, paddle, turbine, etc.). The starch slurry is heated by live steam injection [6] with temperatures of conversion cycle regulated by means of a temperature controller [5]. A typical cycle description follows:

Heat to 158°F (70°C)	15 minutes
Hold at 158-162°F (70-72°C)	20 minutes
Heat to 178°F (81°C)	10 minutes
Hold at 178-182°F (81-83°C)	20 minutes
Heat to 205°F (96°C)	5 minutes
Hold at 205°F (96°C)	30 minutes

Upon completion of the conversion cycle, the temperature of the starch paste can (not standard practice) be regulated by passing cold water [4] through the jacket [7] of the conversion tank and discharging to the drain [8]. The converted starch paste is then removed via the discharge line [9] and transferred to storage.

Construction Materials

The preferred construction material is stainless steel 300 series with exception of steam and water supply.

Items of Special Interest

1. The enzyme generally employed is a bacterial alpha amylase, a proteinaceous structure, which is sensitive to temperature, pH, oxidizing agents, and heavy metals.

2. Enzyme is added at a use level, comensurate to conditions employed and the degree of hydrolysis required, often measured in activity units (SKBs)(1).

3. Starch must be gelatinized to be susceptible to enzyme action.

4. The first temperature plateau 158-162°F (70-72°C) partially gelatinizes the starch and allows the enzyme to hydrolyze the gelatinized portion, thereby reducing peak viscosity and consequent energy load on further heating.

5. The second plateau 178-182°F (81-83°C) provides more complete gelatinization of the starch and resents more optimum conditions for enzyme activity.

6. The third plateau 205°F (96°C) provides more complete dispersion and most importantly, the conditions which result in denaturation or inactivation of the enzyme.

7. Carryover of residual active enzyme into storage will result in continued hydrolysis, uncontrolled conditions, and impairment of product quality.

8. Enzyme conversion units as illustrated are often pressurized (10-15 psig) allowing higher temperatures

of 220-230°F (104-110°C) providing even more complete dispersion of starch and assuring more rapid and complete inactivation of the enzyme.

Paste, Storage, and Handling

An essential factor regarding starch use in a surface sizing application is maintenance of high quality paste or starch dispersion. The design of equipment and conditions maintained are an integral part of achieving maximum effectiveness and high quality.

Starch paste which has been prepared by any of the foregoing methods is collected in a storage tank which is temperature controlled by means of electrical heat tracing. Recommended temperatures are: greater than 15 percent dry solids, 190°F (88°C) or above; less than 15 percent dry solids, 150-160°F (65-71°C).

Figure 4 illustrates a storage and handling system which incorporates many of the features required for proper maintenance. The process in Figure 4 is described as follows:

Starch paste is transferred by means of a centrifugal pump [2] to a run tank [3A] [3B] with recycle returning to storage tank. Enroute water [4] is added by ratio control to obtain the desired application solids. Temperature is maintained in the run tank by means of live steam injection [5]. Starch paste from the run tank is delivered by means of a centrifugal pump [6A] [6B] through an in-line screening device [7] to the size press [8]. Overflow from the size press [9] is returned by gravity to the run tank.

Construction Materials

The preferred construction material for equipment in the storage and handling system is stainless steel 300 series. The exception is equipment used in handling steam and water.

Items of Special Interest

1. The pH of starch should be adjusted (recommended range 8.0 ± 0.5 pH) prior to or immediately upon addition to storage tank.

2. Temperature control in the storage tank by electric heat tracing is effective because the only heat loss is by conduction through tank wall. It is also desirable because it avoids dilution associated with steam injection.

3. Water for dilution should be of good quality (low hardness) and preheated to avoid thermal shock.

4. Gradual cooling or cyclic temperatures should be avoided by using electrical heat tracing and insulation on piping.

5. Starch is highly susceptible to microbial hydrolysis and easily contaminated by airborne organisms. Organisms, particularly thermophilic bacteria, are active at quite high temperatures and require selection of a suitable biocide.

6. Two run tanks are utilized to allow for changeover and cleanup without shutdown.

7. Run tank volume is small to minimize loss and maximize quality.

8. In-line screening is essential to remove extraneous debris including retrograded starch, protein, fibers, etc.

9. Live steam injection is necessary to control temperature in the run tank due to significant heat loss in the size-press cycle.

10. All tanks should be covered to minimize heat loss and contamination.

Controlling pH throughout the system is essential. If pH is dropping in the run tanks, the cause should be determined and corrected or compensated. The most common cause of pH drop is microbial contamination. On occasion, leaching from sheet may affect size solution pH. When leaching is encountered, the only option is additional pH control.

Safety and Precautions

Starches do not normally represent a health hazard. Exposure to eyes or inhalation are generally considered harmful only under unusual or overwhelming dosage. Ingestion of unmodified starch is generally considered harmless. However, it should be noted that the asceptic conditions observed for food starches are not necessarily maintained in processing industrial starches.

Material safety data sheets specific to the product employed should be obtained from supplier.

Proper health and safety practices and suitable use of safety attire, including safety glasses and dust mask, should be employed when handling starch.

Spills and leaks can be maintained by sweeping the area with water, flushing, or both. Waste should be discarded in closed waste containers and disposed of as solid waste. Disposal must comply with local, state, and federal regulations.

Starch dust is highly explosive at critical concentrations. Good housekeeping practice is the best preventive measure. Do not allow dust to accumulate. Avoid exposure of dry starch to excessive heat (ignition temperature

380°F (193°C) or sparks such as welding, smoking, open motors, flame, electric heaters, etc. (2)

Application and Use

Surface sizing is accomplished by three methods: tub sizing, calender sizing, and the most common, size-press application.

Tub sizing results in saturation of the sheet, is relatively slow, and is an energy-intensive process. Tub sizing has limited use achieving unusual or unique finishes and properties.

Calender applications are applied to reduce dusting, linting or piling, and increase sizing. On occasion, dilute size solutions are used to control curl. Calender application is used more on board products, however, can be used on basically all grades.

Size-press systems of varying design and configuration are encountered. Types of size-press systems include: horizontal, vertical, inclined, gate roll, and VACPLY®[2]. Changes have been generated by the desire to increase speed, facilitate handling the sheet, improve side-to-side uniformity, allow application of higher viscosities and concentrations, control side-to-side application, and reduce penetration or strike-in of size. All of these factors affect or are affected by the quality and properties of the size solution.

Starch applications range up to 12 percent pickup on sheet weight. Low applications (0-1.25 percent) generally provide a light finish. More standard levels (1.25-5.0 percent) generally provide good surface sizing while high levels (5.0-12.0 percent) provide internal strength as well.

Preparation and handling of the size-press solution has been discussed previously.

---

[2]VACPLY® is a registered trademark of Karlstads Mekaniska Werkstad, Karlstad, Sweden.

However, it should be noted here that the goal should be to prepare the product best-suited to meet the objectives (sizing, dusting, ink holdout, strength) while meeting the physical properties (viscosity, solids, temperature) dictated by machine conditions.

Literature Citations

1. Sandstedt, R. M., E. Kneen, and M. J. Blish, *Cereal Chemist* 16:712 (1939).

2. Aldiso, D. F., and F. S. Lai, "Review of Literature Related to Grain Dust Explosions," U. S. D. A. Miscellaneous Publication #1375.

Chapter 7

Pigments and Fillers

by Alan J. Bauch

Introduction

By strict definition, pigments are solid, insoluble coloring materials which are added to papermaking furnish to impart some optical property improvement to the sheet of paper. Filler is a much more general term, referring to any material, usually nonfibrous, added to the fiber furnish of paper (1).

This chapter deals primarily with the common pigmentary fillers used in papermaking today. These pigments are added to the papermaking furnish for more than optical property improvements, however. They are also used to economically replace cellulose fiber and to impart functional sheet properties such as porosity, smoothness, and printability characteristics.

Pigments are generally classified as either white pigments or as colorants. White pigments are commonly used in most segments of paper production to improve sheet brightness, opacity, and shade. Although organic dyes are widely used in producing colored papers, colorant pigments have uses in specialty grades where dyes are not suitable because of fading or leaching problems in the paper's end use.

The selection of white pigments available to the papermaker has historically included a variety of naturally occurring or synthetic minerals and, more recently, some synthetic organic polymers. These pigments are available in a number of physical forms and are handled in the mill in a variety of ways, dependent on the physical characteristics of the pigment, the economics of delivery, and the equipment and storage facilities available at the mill.

Generally, pigments are supplied in three forms: dry form in repulpable paper bags, dry form in bulk deliveries, and in slurry form.

How Pigments Improve Brightness and Opacity

When light strikes the surface of a sheet of paper, three things can happen to it. The light can be transmitted through the paper, it can be absorbed by the paper, or it can be reflected by the paper. Absorption and reflection are enhanced by refraction (bending) of light as it strikes the solid surfaces in the paper's structure. A paper's brightness level is determined by the amount of incoming light that is diffusely reflected by the sheet. The opacity level of this sheet is determined by the amount of incoming light that is prevented from traveling through the sheet by reflection and absorption.

The structure of a sheet of paper is essentially a three-dimensional network of cellulose fibers containing a large volume of air space within it. This network is held together by cellulose bonding at the intersections of these fibers. Reflection and absorption occur at the surfaces of these fibers. As the number of air-fiber surfaces in the fiber network increases, the opacity of the sheet increases. However, in a well-bonded unfilled sheet the number of air-fiber interfaces where light reflection takes place is relatively small and these sheets tend to have low-opacity.

White pigments are added to the furnish in the wet end to increase the amount of light reflection and refraction in the finished sheet, thereby increasing the sheet brightness and opacity. Pigment particles are much smaller than the lengths and diameters of wood fibers and they are retained in the fiber network by adsorption to the fiber surfaces. These pigment particles frequently occupy spaces at the intersections (bonding points) of adjacent fibers, so the fiber network of a filled sheet of paper contains not just air-fiber interfaces, but air-pigment interfaces and pigment-fiber interfaces as well. The increased number and types of these interfaces greatly increases the amount of reflection and refraction in the sheet. The pigment particles also act as fiber spacers, preventing some of the inter-fiber bonding and creating more air spaces and air-fiber interfaces which contribute to light refraction. Mineral fillers have a detrimental effect on sheet strength because of this debonding phenomenon.

The total effect of diffuse reflection and refraction of light is known as light scattering. Anything that promotes light scattering in the sheet increases the brightness and opacity of the sheet. Opacity can also be achieved through the absorption of light in the sheet. Examples of materials that absorb light are dyes and colored pigments, which selectively absorb certain portions of the light spectrum. Black dyes or pigments absorb the greatest amount of light, while white pigments absorb relatively little and achieve opacity through light scattering.

The mechanism of the effects of light scattering and absorption on optical properties is discussed in great detail in the theories of Kubelka and Munk. Many articles have been written on this subject and most of today's theories of paper optics are based on the work of these men.

Selection of Pigments

The selection of a pigment or pigments for use in the wet end is based on the optical and physical specifications of the paper, the optical performance and physical properties of the pigments, and, of course, the costs of the pigments.

Some pigments can provide very economical improvements in grades of paper which require only moderate levels of brightness and opacity, while other pigments may be more suitable for achieving higher levels of brightness and opacity. The optical performance of a pigment is affected by such physical factors as its particle size, refractive index, particle structure and shape, as well as its own brightness level and shade.

Light scattering coefficients, representing the effectiveness of a pigment in developing brightness and opacity, are commonly used as a measuring stick for pigment performance. These coefficients are computed from paper optical properties and pigment content using theories developed by Kubelka and Munk. It is a pretty fair generalization to say that the higher the light scattering coefficient of a pigment, the higher its price will be. It is up to papermakers to determine which pigment or pigments will most economically provide them with the level of optical performance needed for each grade of paper they produce.

Pigments Used in Papermaking

Clay

The white clay commonly used in the paper industry as a filling pigment is composed of the mineral kaolinite and is the product of the weathering of feldspar and granite millions of years ago. Kaolin deposits can be broadly classified as primary or secondary deposits. Primary deposits, such as the deposits found near Cornwall, England are located in the same spot in which they were formed. Secondary deposits, such as those found in central Georgia in the United States, consist of kaolin which was transported and sedimented into layered deposits. The central Georgia kaolin deposits were formed along what was the ocean shoreline millions of years ago.

Other kaolin deposits exist in Australia, Brazil, Czechoslovakia, France, Indonesia, Spain, the U.S.S.R., and the People's Republic of China (2). The most important deposits used by the paper industry are those in central Georgia, England, Brazil, and Australia.

The kaolin deposits in the United States are mined by drag-lines and scrapers in open pit mines. The English deposits are typically mined by directing high-pressure water cannons at the deposits and collecting the kaolin-bearing washoff.

Kaolin clay particles consist of many thin hexagonal plates that are naturally laminated into stacks. The grade structure of kaolin clay reflects the type of processing the clay has undergone, the brightness level of the clay, and the particle size distribution of the clay.

Kaolin processing can be done by a dry process or by a wet process. In the dry process, the clay is separated from particulate impurities by air classification to produce what is referred to as air-floated clay. In the wet process, the crude clay is slurried in water, degritted by sedimentation, and separated into the desired particle size range by centrifugal separators. From this point the clay can be sold as a standard brightness grade or improved by bleaching and other impurity removing beneficiation steps. Delaminated clay, produced by grinding kaolin clay particles to separate the thin hexagonal plates, is a specialty grade sometimes used in paper filling.

Clay is supplied to the papermaker in dry bulk form, packaged in repulpable bags, and in slurry form. Standard filling clays that are wet processed are less expensive in slurry form because the manufacturer does not have to incur the expense of removing all water from the finished product. Clay slurries generally range from 65-70 percent solids, depending on the specific grade of clay.

Calcined Clay

The fine particle size, low abrasion, calcined clay used by the paper industry is produced from specially selected and processed kaolin clays by calcining them at a very high temperature to remove the water molecules bound in their crystalline structure. The kaolin crystalline structure is destroyed in the process of calcination, producing an amorphous, anhydrous pigment with unique structural and optical properties.

Calcined kaolins have a slightly higher refractive index than hydrous (uncalcined) kaolin clays, but this difference is relatively small (1.62 for calcined clay vs. 1.56 for hydrous clays) and does not fully explain the superior optical performance of calcined clay. The fine particle size calcined clays have a bulky agglomerate structure and a narrower particle size distribution than the hydrous kaolin clays used in paper filling and coating. These physical and structural properties of calcined clay combine to produce higher levels of paper brightness and opacities than hydrous kaolin clays. In fact, calcined clay is often used as a titanium dioxide extender pigment because of its high optical efficiency.

Calcined clays are supplied in repulpable paper bags, dry bulk deliveries, and in slurry form. Because calcined clay slurries become dilatent (resistant to flow as shear is applied) at lower solids than uncalcined kaolin clay slurries, the maximum practical solids level for calcined clay slurries is approximately 50-55 percent.

Calcium Carbonates

The term calcium carbonate includes a number of physical forms, both naturally occurring and man-made. The calcium carbonates found in nature are composed of the mineral calcite and are processed into pigmentary grades through crushing, wet or dry grinding, beneficiation, and particle size classification. Calcite commonly occurs in rhomboid and cuboid crystalline forms and is mined from chalk, limestone, and marble deposits. The particle size of natural calcium carbonate is controlled by the degree of grinding. The trend in recent years has been to produce finer particle size products for improved optical performance.

Calcium carbonates are also synthetically produced through a number of precipitation processes in which limestone is calcined to produce quicklime (CaO), which is then slaked to produce hydrated lime $Ca(OH)_2$. This is reacted with carbon dioxide gas or other carbonate sources to produce calcium carbonate. The precipitation process can be controlled to produce a product within a desired particle size range and of a particular

crystalline structure. Three crystalline shapes are commonly produced by the precipitation process: acircular (needle-shaped), barrel-shaped, and a rosette configuration of cigar-shaped crystals.

The brightness, shade, abrasion, and other physical properties of calcium carbonate pigments are greatly affected by the amounts and types of impurities present. Natural ground calcium carbonates on the market today are generally 90 G. E. brightness or higher and have particle size distributions ranging from very coarse, 15 micrometers average particle size to as fine as 0.8 micrometers average particle size.

Precipitated calcium carbonates are typically produced with narrower and finer particle size distributions than natural calcium carbonates. Calcium carbonates are used only in neutral or alkaline pH systems because calcium carbonate dissolves in typical low pH systems.

Natural ground calcium carbonates are supplied in repulpable paper bags, in dry bulk, and in slurries with solids contents as high as 76 percent. Precipitated calcium carbonates are also sold in bagged, bulk, and slurry form, though the maximum practical slurry solids is lower than for natural calcium carbonates.

Titanium Dioxides

Titanium dioxide is a high-brightness, high-whiteness pigment and has a high index of refraction. The pigment is processed to high-purity (over 95 percent TiO_2 by weight) from titanium-bearing ore. Two processes are used today, the chloride process and the sulfate process. Two types of titanium dioxide are supplied to the paper industry: anatase and rutile. Anatase has a density of 3.9 g/cc and a refractive index of 2.5. Rutile has a density of 4.0-4.2 g/cc and a refractive index of 2.8.

The high-refractive indices of anatase and rutile titanium dioxide contribute to their superior light scattering (opacifying) powers. Titanium dioxides are generally regarded as the most effective brightening-opacifying pigments used in papermaking. Not surprisingly, they are also among the most expensive pigments used in papermaking.

Titanium dioxide is supplied in dry and slurry forms. Slurry solids range from 65-72 percent, depending on the specific grade of TiO_2.

Talc

Pure talc is a hydrous magnesium silicate mineral having the formula $3MgO \cdot 4SiO_2 \cdot H_2O$. However, pure talc deposits are rarely found and the impurities associated with talcs vary quite widely. Consequently, there is no universal description of the chemical composition of talcs used in papermaking.

The talcs used in papermaking are platy in particle shape, hydrophobic and organophilic (depending on the impurities content and compositions). Because of their hydrophobic and organophilic behavior, talcs are often used as pitch control agents. In this role, talcs are added to the furnish in small amounts (e.g., 1 percent on fiber weight) to adsorb the resinous pitch from the wood pulp, thus preventing it from being deposited onto the wire or picked out on press felts.

Talcs are also used as optical performance pigments in the wet end. They are seldom used alone and often fall into the category of titanium dioxide extenders. Talcs are sometimes used in heavily colored paper grades to adsorb the organic dyes, thus lessening the two-sidedness often encountered in making these grades.

One major drawback in using talcs in the wet end is that their hydrophobic nature prevents their use in slurry form. High levels of wetting agents and dispersants are required to make down hydrophobic talcs and these chemicals are detrimental to pigment retention on the wire. Due to the nature of their mineral impurities, some talcs are somewhat hydrophilic. This makes them more practical as optical performance pigments but less useful as pitch control agents.

Pure talc is a very soft mineral and is quite non-abrasive. However, the impurities associated with many talcs, such as quartz, mica, and calcite, can be quite abrasive and this must be evaluated when considering the use of talc in the wet end of the paper machine.

Due to the difficulty in preparing and handling talc slurries, talc is most commonly supplied in repulpable bags. In order to be most effective in adsorbing pitch, talc is added in dry form to the furnish in the beater or at some other convenient point early in the stock preparation process.

Low-solids talc slurries can be prepared with vigorous agitation. These slurries can be pumped and metered into the stock prep system with centrifugal pumps and conventional flowmeters.

Alumina Trihydrate

Alumina trihydrate pigments are very high-brightness, high-whiteness pigments. They are produced from bauxite ore by reacting the ore with caustic to form sodium aluminate, then hydrolyzing the sodium aluminate to form an alumina trihydrate precipitate (3). This precipitation can be controlled to produce various particle size and surface area variations. Alumina trihydrates are supplied in dry powder form and in slurries of up to 70 percent solids content.

Amorphous Silicas and Silicate Pigments

This heading includes a number of pigments such as amorphous hydrated silicas and sodium aluminosilicates. These pigments are manufactured by the precipitation of silicates (alone or with other elements) to insoluble, amorphous, pigmentary forms. The physical properties of these pigments can be controlled by altering their chemical processing. The ultimate particle size of these pigments typically ranges from 0.03 to 0.3 micrometers, but these particles tend to form aggregates of 0.3 micrometers and upward. Silicas and silicate pigments tend to have high surface areas and, therefore, high oil absorptivity. They are available in bags, dry bulk, and slurry forms but the maximum slurry solids is usually quite low at 30 percent.

Storage and Handling Equipment

Fillers and pigments are usually shipped in one of three forms: dry in kraft bags (usually 50-pound or 100-pound bags, but larger sizes do exist), in dry bulk form, or in slurry form. All three forms are suitable for shipping and handling filler pigments but there are advantages and disadvantages unique to each form.

Most pigments can be bought packaged in kraft bags. These bags can be added directly to the pulper or the bags can be opened so that the pigment can be added at virtually any point in the stock prep system. No investment in storage tanks or make-down facilities is required, but the addition of bagged material requires a great deal of labor input for truck or rail car unloading and bag handling. The higher price of bagged material compared with dry bulk or slurry reflects the supplier's packaging and labor costs in making bagged shipments.

Dry bulk delivery offers the advantages of a lower price than bagged material. Depending on the physical properties of the material, dry bulk shipments can be unloaded and handled in a number of ways such as air conveying, mechanical conveying, and sparging with water.

Air conveying systems can transport dry bulk material with either positive pressure or vacuum. Air conveying has a tendency to reduce the bulk density of a powdered material because of the breakdown of agglomerated particles and the aeration of the bulk mass. A great deal of dust is generated from the aeration of the dry powder and dust collectors are essential to these systems.

Mechanical conveying includes the use of bucket, belt, and screw systems. Belt conveyers and screw conveyers have a drawback in that their ability to convey dry powders up slopes is directly affected by the angle of repose and flowability of that powder. The angle of repose of a powder is defined as the angle formed between the sides of a freestanding mound of the powder with the horizontal. If the angle of repose of a material is low, it will have a tendency to spread out and flow back when being conveyed up a slope.

Some pigments (such as calcined clay, amorphous silicas, and silicate pigments) can be shipped in dry bulk and slurried in the tank car by the customer before unloading. This slurrying process, known as sparging, requires that the pigment wet out readily and disperse easily when mixed by air agitation in the car. Fresh water is pumped into the bottom of the car containing the dry bulk powder and the powder is allowed to soak for a short period. After soaking, the pigment-water mixture is agitated and churned by bubbling air up from the bottom of the car. This produces a low-solids slurry (typically 30 percent solids or less) which can be unloaded and handled as a slurry delivery.

The customer receives the cost benefits of dry bulk price and freight rates along with the ease of slurry handling in the mill.

Many pigments are manufactured by a wet process and the supplier must dry the product to make bagged or dry bulk shipments, resulting in a higher price for the dry product. Once in the mill, slurries offer the customer greater ease and flexibility in unloading, handling, and storage systems than bag or dry bulk deliveries. The product arrives ready for use and can be quickly unloaded and easily routed to a storage tank where it can be stored more efficiently than either bagged or dry bulk product. For example, a 70 percent solids filler clay slurry can contain up to three times the dry clay weight per cubic foot than the same bagged or dry bulk filler clay. There is a significant investment in pumps, piping, and storage tanks for slurry use, however.

Slurry shipments can be delivered by tank truck or by rail tank car. These cars are typically equipped with slurry discharge valves at the top and bottom of the tank, valves for tank pressurization, and sometimes with air sparging valves for agitating the slurry. Unloading can be accomplished by simple gravity flow through the bottom outlet of the tank car if the mill's storage tank is below the level of the car. Otherwise, the slurry can be pumped out of the car to the storage tank or the tank car can be pressurized so that the slurry is delivered to the storage tank without pumping. This last unloading technique requires only a source of compressed air (30 psi is sufficient) with a large enough flow rate to unload the car in a reasonable time.

Pigment slurries are much simpler to handle in the mill than dry pigments. With proper storage and pumping capabilities, pigment slurry use can be controlled more effectively than dry pigment usage and with less physical effort.

Slurry storage tanks are normally made of tile or steel, although fiber-wound plastic tanks have also been used. For practical reasons, a storage tank should hold at least 1 1/2 times the volume of a slurry shipment. Storage tanks should be equipped with agitators to prevent the slurry from remaining static for long periods of time since all slurries will eventually settle out if left unagitated long enough.

Top-entering agitators are strongly recommended over side-entering agitators because of the seal problems encountered with side-entering agitators. Only gentle agitation is required. Storage tanks should have bottoms sloped toward the discharge and vented covers. Slurry should be delivered to the storage tank through top-entering pipes.

Centrifugal pumps are most widely used for circulating pigment slurries and are recommended for high-solids, low-head applications. Low-speed positive-displacement pumps are recommended for handling high-solids, dilatent slurries in high-head applications because of their lower shear rates. Excessive shear must be avoided, as most all pigment slurries will exhibit some dilatency (thickening under high shear application) at high-solids concentrations.

Pigment manufacturers should be contacted for specific recommendations regarding their products. As a general practice, centrifugal pump speeds and impeller diameters should be designed so that the impeller tip speed does not exceed 3,600 fpm. Impeller diameter should be 1/2-inch or 1 inch less than the maximum size accommodated by the casing. Extra gaskets may be put in split-casing type centrifugal pumps to increase the clearances on the sides of the impeller. Centrifugal pumps with suction sizes less than three inches are not considered good practice when handling pigment slurries.

Stainless steel or plastic piping can be used for pigment slurry circulation. High pipeline velocities should be avoided because of possible problems with dilatency. Three feet-per-second is a good general design velocity. Long radius fittings should be used to improve flow characteristics and all piping should be sloped 1/4-inch to the foot toward a storage or waste outlet to allow draining and flushing of the piping. Slurries left undrained in piping will settle out over time and be quite difficult to remove.

Butterfly valves are recommended in piping larger than two inches in diameter. These valves should be installed so the axis of rotation is horizontal and the lower portion of the disc opens downstream. Ball valves are recommended for their superior flow characteristics in piping two inches in diameter or less. Magnetic flowmeters can be used to meter slurry flows. While most pigment slurries are delivered pre-screened, open screens or pressurized screens can be installed at any convenient point in a slurry system. Fifty mesh or 100 mesh are most commonly used.

Application and Use of Pigments and Fillers

Fillers and pigments can be added to the furnish at practically any point in the stock prep system, provided there is sufficient mixing to ensure good dispersion throughout the furnish prior to sheet formation. In many cases, pigments are the first nonfibrous component added to the furnish and pigment retention and sheet formation are generally enhanced by adding the pigments to the furnish prior to alum addition.

Pigments can be added in dry or slurry form. Pigments received in repulpable bags lend themselves to batch addition to the pulper or to a beater chest, while slurries can easily be added to the furnish batchwise or in a continuous metered flow.

The amount of pigment added to the furnish is usually governed by economics and by the optical and physical requirements of the sheet. When using high-performance pigments, which cost as much or more than the fiber portion of the furnish, it is uneconomical to use more pigment than is required to achieve the targeted optical specification of the sheet. When the cost of the mineral filler is lower than the cost of the fiber portion of the furnish, it is advisable to add as much to the furnish as possible. The limiting factor for filler addition in this case is usually a physical property of the paper that must be maintained.

Since mineral fillers do not bond with fibers or with other minerals, strength properties tend to decrease as filler level increases. Also, the density of the sheet increases as filler level increases because mineral fillers are typically denser than fiber. Some grades of paper contain upwards of 30 percent filler pigment by weight, and since filler retention is generally lower than the retention of fiber, these furnishes contain even higher levels of filler pigment.

The final analysis in evaluating a pigment's performance is a comparison of the optical or physical property improvement in the sheet against the costs involved in using that pigment. Because different pigments can cost significantly more or less than an equivalent weight of fiber, the total cost of the furnish is the truest basis for comparing different pigments and pigment blends.

Literature Citations

1. *Dictionary of Paper*, 3rd edn., New York: American Pulp and Paper Assoc., 1965.

2. Murray, Hayden H., "Paper Coating Pigments," *TAPPI Monograph Series No. 38*, R. W. Hagemeyer, ed., Atlanta, GA: TAPPI PRESS, 1976, pp.76-78.

3. Koenig, James J., "Paper Coating Pigments," *TAPPI Monograph Series No. 38*, R. W. Hagemeyer, ed., Atlanta, GA: TAPPI PRESS, 1976, p. 8.

Chapter 8

Retention Aids, Drainage Aids, and Flocculants

by Frederick Halverson

Introduction

In this chapter, the term "flocculation" will be used to denote the overall process whereby particles in suspension join together to form aggregates. An aggregate thus formed is termed a floc or floccule, and any chemical reagent which enhances the process will be termed a flocculating agent or flocculant (1) (2). Within the floc, individual particles tend to retain their identities but are constrained in motion relative to other particles in the floc. In practice, this aggregation process usually involves particles with a distribution of sizes, including small ones depositing on large ones. It is not restricted to particles of colloidal size.

Although flocculation is defined in terms of formation of aggregates, it is the consequences of that aggregation which constitute performance of flocculants in commercial applications. The various aspects of performance are specified by descriptors which measure a given attribute of the material treated with the flocculant. Among these descriptors are the rate at which a suspension can be filtered, the amount of

filler or fines retained in the fiber web, the rate at which water drains during formation of the web, and, in a more general sense, the rate of production. Within a given industry, flocculants are frequently referred to as aids for a specific descriptor, such as filtration aid, retention aid, or drainage aid (1). In this chapter, the terms aid and flocculant will be used interchangeably, and sometimes the designation aid-flocculant will be used for emphasis.

In order to achieve desired performance, more than one aid may be used, either added at the same time or sequentially at different addition points in the system. They will be referred to as multiple aids or coflocculants. In some cases one coflocculant may exhibit poor performance by itself, but performance of a combination exceeds the sum of the parts (synergism). In this case the one component may also be termed a promoter (3). On the other hand, some flocculants (chemical structures) may exhibit good performance for more than one descriptor. Hence, it is convenient to organize a discussion of aids of importance in the paper industry first regarding types of chemical reagent, and then

Table 1. SRUs

I.D.	Molecular Formula	Charge
1.	$+CH_2CH)_n$ — $C(=O)-NH_2$	nonionic
2.	$+CH_2CH_2O)_n$	nonionic
3.	$+CH_2CH)_n$ — $C(=O)-OH$	anionic
4.	$+CH_2CH)_n$ — (benzene ring) — SO_3H	anionic
5.	$+CH_2CH)_n$ — $C(=O)-NH-C(CH_3)(CH_3)-CH_2-SO_3H$	anionic
6.	$+CH_2CH)_n$ — $C(=O)-OCH_2CH_2N(CH_3)(CH_3)$	cationic
7.	$+CH_2C(CH_3))_n$ — $C(=O)-OCH_2CH_2N(CH_3)(CH_3)$	cationic
8.	$+CH_2CH)_n$ — $C(=O)-OCH_2CH_2N(CH_2CH_3)(CH_2CH_3)$	cationic
9.	$+CH_2C(CH_3))_n$ — $C(=O)-NHCH_2CH_2CH_2N(CH_3)(CH_3)$	cationic
10.	$+CH_2C(H))_n$ — $C(=O)-NHCH_2CH_2CH_2N(CH_3)(CH_3)$	cationic
11.	$+CH_2CH)_n$ — $C(=O)-NHCH_2N(CH_3)(CH_3)$	cationic
12.	$+CH_2$——$CH_2)_n$ — ring $N^+(CH_3)(CH_3)$ Cl^-	cationic
13.	$+CH_2CH(OH)CH_2N^+(CH_3))_n$ Cl^-	cationic
14.	$+CH_2CH_2N(H))_n$	cationic
15.	$+CH_2CH_2N)_n$	cationic
16.	$+CH_2CH_2-NH_2$	cationic

proceed to how they are used in specific areas.

Aids-flocculants listed in this chapter are representative of the major types reported as used in the industry. In practical terms a distinction should be drawn between an aid-flocculant type and the specific commercial product which contains it, since product formulation will be different between suppliers. Thus, there can be differences in stability, ease of dissolution, and performance between products which appear to contain the same basic aid-flocculant. Suppliers are not identified in this chapter. Application techniques and uses are representative, rather than exhaustive. In keeping with recent publication policies, SI units (4) are used, but for clarity more conventional units are also included.

Chemical Composition of Aid-Flocculant Types

Insofar as chemical origin is concerned, aids can be grouped into three major categories: synthetic organic polymers or polyelectrolytes, natural products and their derivatives, and inorganic materials. Representative structures of utility in the paper industry will be indicated in each category with brief comments (1).

Synthetic Organic Polymers

Practical synthetic organic aids are water-soluble polymers with weight average molecular weights ranging from tens of thousands to tens of millions g/mol. If none of the subunits in the polymer molecule carry a charge the polymer is termed nonionic. If some of the subunits are charged, it is called a polyelectrolyte; termed cationic if the charge is positive, anionic if the charge is negative, ampholytic if both positively and negatively charged subunits are present in the same polymer chain. In the following discussion

the structures are grouped as nonionic, anionic, cationic, and ampholytic with some additional subdivisions (1) (5-7).

Due to space considerations, representative structural repeating units (SRUs) occurring in a variety of commercial flocculants are depicted schematically in Table 1 (based on Table 2 in reference 1). A single polymer molecule may contain one, two, three, or more different SRUs (homopolymer, copolymer, terpolymer). The I. D. column in Table 1 is an identification number to reference the structure indicated. When there are two or more SRUs in a polymer chain, the possibility for some compositional heterogeneties exists. For example, there may be a variation in the relative numbers of the different SRUs between copolymer molecules within the product, giving rise to a chemical composition distribution (CCD) on a microscale (8) (9). There also is the possibility for variations in the sequence distribution (SD) of the SRUs along individual polymer chains (10-12). Neither of these microscale phenomena are reflected in data on average values of chemical composition for a product, but may be reflected in performance characteristics, since suspension particles see one molecule at a time (13) (14). Synthesis procedures and purity of reactants can influence parameters.

In the following discussion, SRUs and the monomers which give rise to them sometimes will be used interchangeably. Also, instead of showing separate SRUs for acid and base forms, only the acid form is sketched for anionic SRUs and the base form for cationic SRUs.

Synthetic organic aids-flocculants are supplied to the user in a variety of physical forms: dry powder, granules, aqueous solutions, gel logs, and emulsions. Emulsions are of the type described as inverse emulsions, that is water-in-oil, where aqueous droplets (essentially gel particles) are

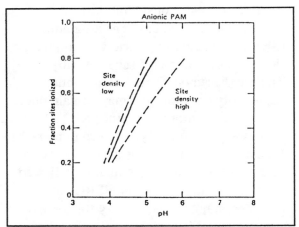

Figure 1.   Carboxyl Ionization in Acrylic Acid-acrylamide Copolymers
Courtesy of Canadian Mineral Processors (17)

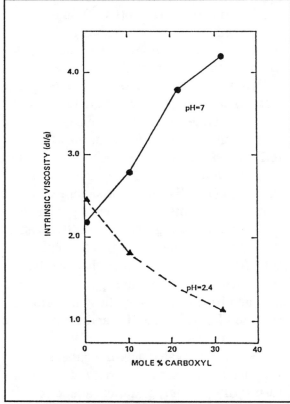

Figure 2.   Molecular Variation with Changing Carboxyl Level
Courtesy of American Cyanamid Company (18)

dispersed in a continuous oil phase. The current standard emulsion has droplets 1-5 micrometers in diameter, which in the following will be referred to simply as an emulsion polymer. Common emulsions contain about 30 wt% active polymer, but higher levels are available (15). Not all aids-flocculants are available in all forms.

Nonionic

It is important to note that polymers sometimes are classified as nonionic even if they contain a low level of charged subunits. One reason is the difficulty in quantitatively measuring very low levels of charge in a polymer molecule and so a practical cut-off point of <1 mol% seems reasonable.

Poly(acrylamide). Acrylamide homopolymer (SRU #l) can be obtained over a wide range of molecular weights, up to 20 million g/mol. In the synthesis, particularly for high molecular weights, care must be exercised to avoid adventitious hydrolysis of amide, thereby introducing some carboxyl groups (SRU #3). Some manufacturers appear to allow up to several percent in their nonionic products, but it can be controlled to a few tenths of one percent.

Although one would expect the term polyacrylamide to have the same meaning as poly (acrylamide), common usage has given it a broader meaning; generally, any polymer containing ≥50 percent mol% acrylamide subunits in the chain. Within this scheme of usage, it is convenient to designate such an acrylamide-based polymer by the term PAM and use the prefixes N, C, or A to designate the nonionic, cationic, or anionic species when desired.

Poly(ethylene oxide). The molecular weight for flocculant-grade ethylene oxide polymers, SRU #2, is considerably higher than for other applications, generally in excess of three million g/mol. This polymer is frequently referred to as PEO.

Anionic

Several types of anionic sites in a polymer molecule are possible, but only two are of commercial significance.

Polymers Containing Carboxyl Groups. The homopolymer of acrylic acid, SRU #3, and copolymers with acrylamide, SRU #1, dominate the anionic market. Poly(acrylamide) can be hydrolyzed readily with alkali to introduce acrylic acid sites along the polymer chain, up to about 50 percent conversion of the amide groups. Copolymerization of acrylic acid with acrylamide covers the entire anionicity range. There can be some variation in the sequence distribution of carboxyl groups along the chain depending on how the A-PAM is synthesized (10) (11) (16).

Although the carboxyl grouping is a moderately strong acid, the extent of ionization (hence number of anionic sites) is a function of pH in acidic media. There also is an effect of neighboring groups, and of the ionic strength of the medium. A representative situation is plotted in Figure 1 where the solid line represents an acrylic acid-acrylamide copolymer with 30 mol% acrylic acid, in a medium of moderate ionic strength (17).

Another feature which enters in acidic media for an A-PAM of this type is the tendency for unionized carboxyl groups to coordinate with each other and collapse the molecule. An extreme example is illustrated in Figure 2 where intrinsic viscosity is plotted against anionicity for two solvent systems of comparable ionic strength, one acidic and the other neutral (18). The polymer molecules all have the same degree of polymerization, with a molecular weight of about 600,000 g/mol. Since the intrinsic viscosity is roughly

proportional to the effective size of the molecule in in the solvent, this plot indicates how the molecule expands or contracts with pH as the carboxyl level is changed.

Polymers Containing Sulfonic Acid Groups. The sulfonic acid grouping is an intrinsically stronger acid than the carboxyl grouping, hence it remains essentially completely ionized in media at a low pH. Available products contain either styrenesulfonic acid, SRU #4, or 2-acrylamido-2-methylpropanesulfonic acid (AMPS), SRU #5. They may be homopolymers or copolymers with acrylamide. The latter can be prepared with a wide range of anionicity and molecular weight.

Cationic

Although other possibilities exist, tetravalent nitrogen appears to be the charged site in all commercially available cationic flocculants at the present time. The charge is derived from protonation of amine groupings or from quaternary nitrogen salts. The former are pH-dependent whereas the latter are not. Fractions of potential cationic sites (amines) which actually carry a charge as a function of pH are shown in Figure 3 for four representative polymer types. The polymer represented by curve A contains only quaternary nitrogen; the Mannich reaction product of poly(acrylamide), represented by curve D, contains tertiary nitrogen groups pendant to the polymer chain; the poly(ethyleneimine), represented by curve C, contains a distribution of primary, secondary, and tertiary amine groups; and the polymer represented by curve B has a special resonant structure which makes it particularly basic (amidine type), but it is not commercially available.

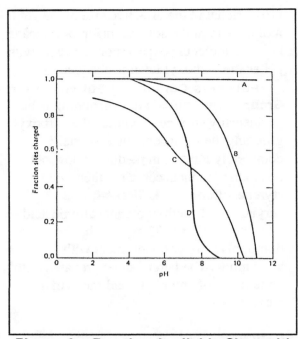

Figure 3.   Fraction Available Sites with Cationic Charge
(A) Quaternary nitrogen, poly (N, N, N-tri-methylallylammonium methyl sulfate); (B) poly(vinylimidazoline); (C) poly(ethyleneimine), Montrek 1000; (D) Mannich reaction product of poly(acrylamide). Reprinted from reference (1) with permission.

Dialkylamino alkyl Acrylic and Methacrylic Ester Polymers. The N-(dialkylamino alkyl) acrylic ester polymers, SRU #6 and SRU #8, and N-(dialkylamino alkyl) methacrylic ester polymer, SRU #7, both as homopolymers and as copolymers with acrylamide (SRU #1), represent an important class of cationic polymers. Of major importance are the methyl chloride and dimethyl sulfate quaternary salts. As derivatives of acrylic acid and methacrylic acid, the amino ester monomers copolymerize readily with acrylamide to yield high molecular weight polymers over a wide range of cationicities.

N-(dialkylamino alkyl) Acrylamide Polymers. These are similar to the acrylic and methacrylic acid ester polymers. Polymers with SRU #9 and SRU #1 are commercially available primarily as the quaternary salts. Since they are derivatives of acrylamide, the monomers copolymerize readily with acrylamide (SRU #1) to yield high molecular weight polymers and can be made with a wide range of cationicities.

Cationic Carbamoyl Polymers. The homopolymer of acrylamide reacts with formaldehyde and dimethylamine via the Mannich reaction to form a polytertiary amine, SRU #11. Starting with N-PAM of the appropriate molecular weight, highly cationic polymers can be prepared (essentially a copolymer of SRU #11 with SRU #1). As a tertiary amine, cationicity is a function of pH (see Figure 3). The polymer is available as aqueous solutions, but a combination of high molecular weight and high cationicity mandates low polymer solids (< 10 wt%) to keep viscosity in a practical range for handling.

Poly(diallyldimethylammonium Chloride). This polymer, SRU #12, can be prepared in the low to medium molecular weight range ($<10^6$ g/mol), and as a quaternary salt it is insensitive to pH variations. Copolymers with acrylamide (SRU #1) also are of importance and can be prepared with a wide range of molecular weights. The monomer diallyldimethylammonium chloride is frequently abbreviated as DADM.

Poly(hydroxyalkylene polyamines). A very important commercial polymer in this group is the quaternary salt of poly(2-hydroxypropyl-1-N-methylamine), SRU #13. Note that the charge resides on a nitrogen atom in the backbone of the polymer chain, contrary to the previously listed structures in which the charge is on a pendent group. Cationic polymers of this type are termed ionenes. Polyamine quaternaries are sometimes called "polyquats." Molecular weights of these polyquats generally are below 10^5 g/mol. The molecular weight can be increased by addition of small amounts of a cross-linking agent, such as ethylene diamine, which appears to give rise to a highly branched spherical star type of molecule (19).

Poly(ethyleneimine). This polymer is produced commercially from ethyleneimine, is frequently designated as PEI, and is highly branched with primary (SRU #16), secondary (SRU #14), and tertiary (SRU #15) amine groupings. There is a significant amount of each type of amine present (20) and the pH dependency of cationicity for one commercial sample is shown in Figure 3. Modified polyethyleneimines with molecular weights up to 2 times 10^6 g/mol are also available (21).

Cationic Cyanamide Derivatives. A diverse group of amino-aldehyde step-growth condensation products have been reported, with amino building blocks which could be considered as derived from cyanamide as a basic raw material (1). Formaldehyde is used as the aldehyde, and the more common amine building blocks are melamine, urea, and dicyandiamide. The molecular weight of these products does not appear to be high (22). They are water-dispersible and may be modified with multifunctional amines and amino alcohols.

Ampholytic

Various polyampholytes are mentioned in the literature as flocculants. Most can be considered terpolymers incorporating SRUs from Table 1. Acrylic acid (SRU #3) is a common component together with a cationic amine, and frequently acrylamide (SRU #1). An example is a terpolymer synthesized from acrylic acid, DADM, and acrylamide (SRU #3-SRU #12-SRU #1). Other polyampholytes include SRU #3-SRU #11-SRU #1 and quaternary salt of SRU #7-SRU #3 (23). These polyampholyte also are

amphoteric, that is, can function either as acid or base. Synthetic polyampholytes do not appear to have established a significant commercial position.

Natural Product Derivatives

Starch

Starch consists of two types of polysaccharides: linear chains of D-glucose units joined by ∝-D(1 → 4) linkages (amylose), and highly branched structures consisting of short chains of 20-25 units of the linear polysaccharide joined to a central chain by ∝-(1 → 6)) glycosidic linkages (amylopectin). Molecular weight of the amylose ranges from 10,000 to 50,000 g/mol, while the molecular weight of amylopectin may range up to many million g/mol (1) (24) (25).

Starch is derived from two main plant origins, cereal and root, such as corn, maize, rice, wheat, tapioca, and potato. Starch granules are isolated from the raw material by a wet milling operation, refined, and dried (26). During the final phases of the refining process and prior to drying, the cleaned, unmodified starch slurry can be diverted to any of several treating processes, such as derivatization (introduction of cationicity).

In all these stages of treatment the starch is maintained in the granular form. The polymers are released from granular form by heating an aqueous slurry (cooking), whereby the granules disintegrate. There is a difference in the percentages of amylose and amylopectin in commonly available starches, with waxy maize containing almost 100 percent amylopectin compared to about 80 percent for potato starch (26) (27). Amylose and amylopectin are nonionic polymers.

Cationic Starch Derivatives. These are produced by etherification with reagents such as N, N-diethylaminoethyl chloride and 2, 3-epoxypropyl trimethylammonium chloride.

Degree of substitution (DS) is generally low (<0.05) but higher levels are available.

Anionic Starch. This can be produced by causticizatian with alkali or by treatment with mono- or disodium phosphate at elevated temperatures.

Amphoteric Starch. This can be produced by giving cationic starch a second treatment to introduce anionic sites (28). DS is low for practical products.

Guar Gum

This is a water-soluble, high molecular weight, nonionic polysaccharide (114). It is a legume seed extract, but mechanical seeding and harvesting have made it available on a large scale. Chemically it is poly-$(1 \rightarrow 4)$-ß-D-mannose substituted on mannose units with D-galactose through $\propto (1 \rightarrow 6)$ links. Substitution is somewhat random along the chain, with about half of the mannose units having a galactose attached. It has the ability to bind to cellulose fibers. Nonionic guars modified by introduction of groups such as hydroxyethyl are also available.

Cationic Guar. This is obtained via aminoethyl or quaternary ammonium derivatives with DS <0.1.

Anionic Guar. This is obtained by the introduction of carboxyl groups with DS <0.1.

Inorganic Products

Aluminum Derivatives

Aluminum is a hydrolyzing cation and a common description denotes hydroxide ions as explicit ligands, with all others implicit, as in the generalized reaction scheme:

$$x\ Al^{+3} + y(OH)^- \rightarrow Al_x(OH)_y{}^{(3x-y)+} \quad (29)$$

There have been many investigations of hydrolysis which differ in detail, but there seems to be a consensus on general features (30-32). At low and high pH, the main species are monomeric, namely Al^{+3}, and $Al(OH)^{+2}$ at the low end, and $Al(OH)_4{}^-$ at the high end. Slow addition of base to aluminum chloride or nitrate solutions of moderate concentration can yield clear solutions up to an OH/AL ratio R of nearly 2.7, with the formation of polymeric species. Beyond this point, precipitation of hydroxide occurs. The polymeric species are metastable systems which eventually will precipitate as crystalline hydroxides, although they may not do so for a year or more. When diluted, they will slowly depolymerize. The most stable polymer cation is $Al_{13}(OH)_{32}{}^{+7}$, but a more descriptive formula with all ligands and with chloride or nitrate counterions follows:

$$[AlO_4\ Al_{12}\ (OH)_{24}\ (H_2O)_{12}]^{+7}$$

Investigations on the molar volume of the cation indicate it is a compact entity behaving as a solid microparticle with a diameter of ca.1.5 nm (33).

Sulfate ions appear to act as a catalyst in the precipitation process. With sulfate present, a visible precipitate can be observed at smaller R values, depending on the concentration of aluminum and sulfate (32b). Generally, these studies have been made at higher levels of aluminum and sulfate than commonly encountered in papermaking (34). Furthermore, hydrolysis reactions are not instantaneous, and so the time scale under conditions of use becomes a factor (106).

Alum. Since alum is reviewed in Chapter 4, comments here are brief. Papermaker's alum refers to a commercial aluminum sulfate hydrate, $Al_2(SO_4)_3 \cdot XH_2O$, where X is about 14. It is available either in dry form or as an aqueous solution. Dry alum is available in several grades, with a minimum aluminum content of 17 wt% expressed as Al_2O_3. Liquid alum is about a 49 percent solution of aluminum sulfate hydrate, or about 8.3 wt%

aluminum as Al_2O_3. Alum is widely used in papermaking for a variety of purposes and frequently is a de facto component for drainage and retention (35) (36). It is also widely used in water treatment (37).

Sodium Aluminate. Sodium aluminate is a white crystalline solid having the empirical formula:

$NaAlO_2$ or $[NaAl(OH)_4]$

It provides a strongly alkaline source of water-soluble aluminum, particularly useful when addition of sulfate ions is undesirable. It is commercially available either in dry form or in solution with an excess of base present. In papermaking it sometimes serves as partial replacement for alum (38).

Polyaluminum Compounds. A partially hydrolyzed aluminum chloride solution with R ca.1.5, known as poly(aluminum chloride), or PAC, was introduced in Japan as an alternative to alum (39) (40). It usually contained some sulfate. An approximate empirical formula is:

$Al(OH)_{1.5}(SO_4)._{125}Cl_{1.25}$

Stabilizers and related materials (41) have been described in the patent literature and a number of formulations are available commercially. Uses in papermaking have been reported. (42) (43) (109).

Silica

Activated Silica. This material has an early history as a flocculant prepared by acidification of sodium silicate solutions (44) (45). Solubility of amorphous SiO_2 in equilibrium with orthosilicic acid, $Si(OH)_4$, is low, below pH 9, but in more alkaline systems soluble anionic silicate species form. When the pH of an alkaline sodium silicate solution of moderate concentration is lowered into the insolubility range, polymerization occurs. This gives rise to metastable polysilicates or colloidal aggregates carrying negative charges which can be diluted and still retain activity for significant periods of time. Polymerization appears to give nuclei 1-2 nm in size (46), and further growth in the presence of salt appears to be a combination of particle size and of particle aggregation to an open network type of structure.

Colloidal Silica. The term colloidal silica has come to mean a sol of nonporous silica particles (47). Development of the latter requires reduction of the salt level to stabilize the 1-2 nm nuclei against aggregation, and thus allow particle growth by molecular deposition. Details of the growth mechanism point to a uniform particle size in a stable sol. Commercially available sols span the particle size range from several nm to over 100 nm.

Although both activated silica and colloidal silica are flocculants, the higher cost of the latter means it is considered only when the nonporous nature of the particle or the uniform particle size is important. Surfaces of the particles can be modified by various treatments, including converting the normal negative charge of silica to a positive charge (48).

Storage

Typical conditions will be mentioned subject to the proviso that manufacturer recommendations be observed.

Synthetic Organic (Polyelectrolytes)

Commercial materials are available as dry powders or granules, aqueous solutions, and emulsions. The dry materials are usually shipped in plastic or paper bags, or plastic-lined fiber drums, although larger containers can be used where handling facilities are available. Hygroscopicity of the powders must be taken into account. They should be stored in a cool, dry area protected from sunlight. Storage life is a function of product, but generally runs months to years (49).

Aqueous solutions have viscosities that range from very low, as with water, to a thick syrup. They are available in 18.9L (5 gallon) pails, lined drums, tank trucks, and railroad tank cars. Bulk storage is common using fiberglass tanks or lined metal tanks. It is desirable to protect the product from temperature extremes. Whereas freezing probably doesn't harm the aqueous solutions, it may be inconvenient to thaw satisfactorily on demand. Again, storage life is a function of the product with times ranging from a month to a year.

Storage of emulsions is similar to aqueous solutions. However, due to possible settling of droplets during long standing, it is desirable to provide storage systems with some type of agitation. It also is desirable to store in the temperature range of 41-87°F (5-30°C).

Natural Products

One difference between synthetic polymers and natural product derivatives is that the latter generally are more susceptible to biodegradation which may limit storage time. Otherwise, storage and handling are similar.

Inorganic Products

Most inorganic products allow prolonged storage longer than needed for prudent inventory. Forms generated on-site, however, such as activated silica, normally do not have the same storage stability and should be used within hours.

Safety Data

Manufacturers are required to provide an MSDS with each product and procedures recommended therein should be followed.

Most polymeric flocculant molecules themselves appear to exhibit low mammalian toxicity, but the same is not necessarily true for the corresponding monomeric species. An example is polyacrylamide compared with acrylamide monomer (50) (51). Production procedures, however, can hold residual monomer in the flocculant product to a very low level. Purity requirements for a product used on paper fibers in contact with food is greater than for a product used on tailings in mining operations, and may require additional steps in manufacturing.

Safety in the workplace also means good practice in industrial handling. Aqueous solutions of water-soluble polymers tend to be slippery. Spills should be avoided and cleaned up promptly if they occur.

Application Procedures and Product Uses

Because of the variety of flocculants available and the many ways in which they are handled, a detailed discussion of all aspects is outside the confines of this chapter. Instead, typical features will be mentioned along with the proviso that supplier recommendations be observed.

Make-up

The process of converting the product as received from the supplier into a dilute aqueous solution for addition to the substrate being treated will be referred to as make-up (49). The water used in the process will be termed make-up water. Procedures in make-up and composition of the make-up water can be a significant aspect of performance by an aid. Factors such as pH, ionic strength, presence of multivalent ions, and temperature can influence flocculants in different ways. For example, the presence of ferrous ions can cause significant molecular weight degradation of acrylamide-based polymers upon standing in the presence of air (52) (53). It is this type of problem which usually can be avoided by noting the manufacturer recommendations.

For many flocculants (particularly higher molecular weight products) make-up is carried out in two stages. An initial stage develops a moderately viscous aqueous solution which is an intermediate on the way to preparing the dilute solution (second stage) appropriate for addition to the process stream. This approach allows convenient handling to obtain good solutions and enhances metering of the flocculant.

Aqueous Products

Preparation of dilute solutions of aqueous products is essentially the second stage of make-up as mentioned previously. If the product viscosity is very low, dilution can be straightforward and rapid. On the other hand, if the product viscosity is high, one must always keep in mind that mixing with water does not provide an instantaneously uniform molecular dispersion (54). Mechanical mixing can provide small-scale ribbons or packets of the viscous material (55) (56), but molecular dispersion relies on Brownian motion which requires time. Furthermore, allowable intensity of mechanical mixing is conditioned by shear stability of the flocculant (57) (58). In practice, a compromise is made between intensity of mixing and time, vs. aid performance, which fortunately is compatible with the time scale of plant operations. In most cases, in-line mixers can be used (49) (55). Suppliers usually provide recommendations.

For molecular weights below 10^5 g/mol, mechanical degradation is not a problem. Nor is it a problem with sols, and intensity of mechanical mixing can be high.

Dry Polymeric Products

Both dissolution and dilution are involved for these products. It is critical that the dry flocculant be added to the make-up water in a manner which provides good dispersion of particles (such as via a flocculant disperser) otherwise lumps or "fish-eyes" will form which are difficult to dissolve. If left in the solution they can be a problem later on. Particles of high molecular weight polymer require time to dissolve, and it is necessary to keep them dispersed during this period. Vigorous stirring will accomplish it, but as dissolution occurs the solution becomes more viscous and the possibility of molecular weight degradation increases (57) (58). Thus, an appropriate balance between dispersion and shear stress needs to be attained. For practical reasons, make-up is usually carried out in two stages with dissolution at a concentration allowing the solution to be pumped around without excessive shear, and the dilution required for satisfactory use occuring at a later stage via in-line static mixers (55). The dissolution stage generally is shortest for charged polymers (30-45 minutes) and longer for nonionic high molecular weight polymers (one hour or longer).

The dilution stage is much the same as for aqueous products and requires some time to obtain a molecular dispersion. Indeed there are patent claims that some solutions exhibit better performance if allowed to age before use (59). On the other hand, inventory, storage, and handling have a cost which has to be taken into account when attempting to optimize cost performance for a system.

Emulsions

Some emulsions require the addition of a demulsifier to the make-up water before the emulsion will dissolve in it, but most current products do not. The first step involves dispersing the oil phase rapidly in the make-up water, allowing the small aqueous droplets (gel-like particles) to disperse. This first step can be accomplished with high shear mixing without any effect on the polymer. During the second step, the small gel particles dissolve best with mild agitation. Since they are much smaller than the powder particles they dissolve more rapidly (5-15 minutes). In order for these solutions to be pumped and handled, concentrations are usually at 2 percent or less. The solutions are turbid due to the dispersed oil droplets.

Dilution of this viscous solution to the low concentration appropriate for addition to the process stream is the same as for aqueous products. Dilute solutions for most high molecular weight polymers should be used within 24 hours.

Polysaccharides

Starch is generally shipped to the user in dry granular form. If it is an unmodified starch, various treatments are necessary at the mill to control viscosity and to introduce functionality as desired. On the other hand, starches can be purchased in a preconverted form having been treated to control viscosity and alter the chemistry of the starch molecules (derivatization). These are supplied with instructions for cooking to get the best results. This generally involves dispersing the starch granules in water at the desired concentration (about 3-5 wt%) with agitation and allowing time for complete wetting of the starch granules. The temperature is then raised to 194-203°F (90-95°C) to cause swelling, then subjecting granules to shear with consequent break up into individual starch molecules. This can be accomplished in a batch operation using steam and high-speed mixers. Continuous jet cookers mix the starch slurry with steam in a venturi jet, getting simultaneous heating, swelling, and

shear to break up the granules (60). Temperature profiles are important. The cooking operation is carried out at a higher solids level than appropriate for addition to the stock, and so a dilution step is necessary.

Solutions of pregelatinized starch and guar products are made up much the same as for the dry polymeric products described previously.

Inorganics

Since the inorganics do not involve high molecular weight polymers with random coil structures in solution, dissolution and dilution generally are faster. Those which are supplied as an aqueous product can be diluted easily.

Mechanism of Action

It is clear that in order to function, the aid-flocculant molecules must contact substrate particles and interact with them. In general, the surfaces of the particles carry negative charges which establish a cationic demand (CD) for charge neutralization, a demand which must be satisfied to some degree to allow particle-fiber or particle-particle deposition (61). Thus cationic aids come to mind first, but the furnish usually is a complex mixture with each component influenced by the presence of others and multiple component systems having advantages. As a consequence, anionic and nonionic aids have their utility. It is clear that aids should be added to the system as a dilute solution and should be well-dispersed among the components as quickly as practical. This chapter is directed toward the types of aids-flocculants available and how they are used. A discussion of mechanisms of flocculant action is outside the scope. Fortunately, rather extensive reviews of various theories pertinent to the paper industry are available (62) (63).

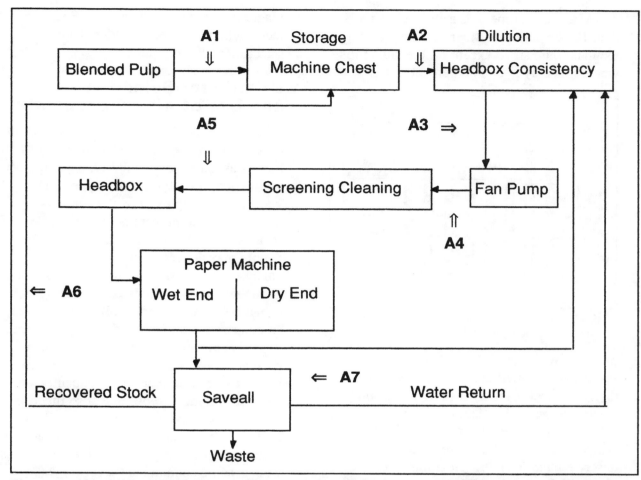

Figure 4. Simplified Schematic of Papermaking Stages for Aids Addition Points (A1 - A7)

Addition Points

Performance of aids-flocculants is dependent not only on the chemistry and physics of surface interactions, but also on hydrodynamic details of the operating system. Where does aggregation occur? What type of environment do the flocs see as they traverse the system? How long are they subjected to various forces ? The strength of interaction between an aid and the surface of a fiber (which binds two surfaces or particles together) relative to the mechanical forces which the fiber or particle may encounter during pulp flow through the papermaking operation, is a crucial factor in performance of an aid (64) (65).

Where should aids be added to the system to be most effective?

For convenient reference, a simplified schematic block diagram of a hypothetical papermaking wet-end system is detailed in Figure 4. In actual practice, some of these stages may be combined or others added. Potential points for addition of aids-flocculants are indicated by AX, where X is a numbered point of addition. Estimates of maximum hydrodynamic shear stresses on fiber surfaces for various wet-end components in papermaking are given in Figure 5 (from reference 66) for a paper machine operating at 760 m/min (2500 ft/min).

At the same time, it is estimated that less than 10 percent of the fibers are subjected to the maximum stress except for the table rolls where this may rise to 50 percent. Large

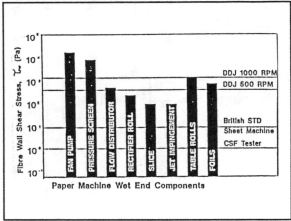

Figure 5.   Reprinted from reference (<u>66</u>) with permission.

particles are more easily dislodged from surfaces than small ones. Since flocs have different levels of resistance to disruption by shear stresses and different responses to reflocculation after disruption, terms such as soft and hard flocs have been used to describe them (<u>67</u>). It is clear that many places in the paper machine have sufficient forces to break up flocs down to a primary flocculation level. Some wet-end components have large enough stresses to fracture polymer molecules (<u>68</u>). The shear stress increases as machine speed increases, which underlines the importance of considering the entire system of aids and hydrodynamics (<u>66</u>) (<u>69</u>).

Since the consistency of the stock is 0.5 percent or higher, it is desirable that flocculant molecules be distributed throughout the slurry at the feedpoint as quickly as possible to allow uniform interaction. This may involve some type of distributor head in the slurry, or introduction at a station exhibiting strong turbulence. It is obvious that flocs formed at Point A1 will be subject to more disruptive forces and for a

longer time than flocs formed at Point A5, a factor which must be considered when selecting addition points.

Criteria for Selection and Use of Aids-Flocculants

Overall criteria for use of aids-flocculants depend on cost performance, including such factors as machine runnability, paper quality, production rate, stock recovery, etc. The variety and complexity of stock and papermaking equipment preclude simple recipes for optimum use. Enhancing one feature may detract from another and a balance of needs must be met. Some general comments on use of aids-flocculants to enhance selected operational parameters are summarized here. It should be kept in mind that all components of the stock, whether present naturally or deliberately added, can interact with each other. Consequently, an additive for one purpose may also function as a de facto additive for another (<u>70</u>-<u>72</u>). Also, as the reuse of water is increased (closing the system) the electrolyte level increases (<u>73</u>) as well as some contaminant levels (<u>74</u>) (<u>75</u>).

Although many reports on performance of aids-flocculants have appeared in the literature, detailed physicochemical composition information on commercial aids is seldom available. Where a competitive situation exists, voluntary release of proprietary information is precluded, although each manufacturer probably has considerable information on a competitive product. Instead, the literature contains qualitative terms such as low, medium, or high as applied to cationicity, anionicity, and molecular weight, or unspecified cationic monomer. On the basis of patent disclosures, of monomer costs, handling, and availability, and of other sources of public information, the following comments appear reasonable. In terms of ranges, level of ionicity within a polymer can be categorized as the percentage of SRUs capable of providing a charge as follows:

Low	1-10	percent
Medium	10-40	percent
High	40-80	percent
Very high	80-100	percent

Similarly, average molecular weight can be lumped into categories as follows:

Low	10^3-10^5	g/mol
Medium	10^5-10^6	g/mol
High	1-5×10^6	g/mol
Very high	$>5 \times 10^6$	g/mol

Unless data are presented on how molecular weights were determined, some of the very high molecular weight values reported in the literature should be accepted with caution.

The most common commercial high molecular weight C-PAM products probably involve the quaternary of SRUs #6 or #7 as the cationic moiety, and fall in the low or medium ionicity category. Since the cost of cationic monomers is considerably higher than acrylamide, the level of cationicity probably tends toward the middle of the low range and toward the lower portion of the medium range. The most common low molecular weight, very high-cationicity products probably involve polyquats (SRU #13), poly(DADM) (SRU #12), or polyethyleneimine (PEI). The effective cationicity of PEI, of course, is pH-dependent (see Figure 3). From a cost factor, a high molecular weight, high cationicity product probably involves SRU #11, the Mannich derivative of N-PAM, or a modified PEI, both of which are pH-dependent (see Figure 3).

As mentioned previously, anionic products are dominated by copolymers of acrylic acid (SRUs #3 and #1), and generally are used as high or very high molecular weight products. Anionicities commonly used in the paper area appear to fall in the low and medium categories. Effective anionicity is pH-dependent on the acid side (see Figure 1). Since introduction of carboxyl anionicity is much cheaper than cationicity, wider ranges can be used without a similar cost factor. When an A-PAM product has essentially no pH-dependence on the acid side, it most probably involves SRUs #4 or #5. If it is a high molecular weight product, the anionic moiety most likely is SRU #5 due to synthesis conditions. Again, there will be a tendency to fall in the lower part of the ionicity ranges due to cost factors.

PEO for this type of application is in the high or very high molecular weight category. Ionic natural products generally fall in the low ionicity range, with DP <0.1 corresponding to less than 10 percent ionicity.

Drainage

Wet-end dewatering, or drainage, is a key factor in runnability of a paper machine, including speed at which it can be operated. Water is frequently considered to be present in two modes: as "free water" which can be removed mechanically, and as "imbibed water" which requires heat ([76]) ([77]). A drainage aid should facilitate removal of free water. "Water removal on a fourdrinier table could be ideally divided into three stages: (a) gravity drainage from the slice to the first set of foils, (b) low-vacuum drainage in the foils and table roll area, and (c) high-vacuum dewatering from the first flat boxes to the couch" ([77]).

A paper machine is a dynamic system allowing limited time in each stage.

Dewatering in the early stages is a slurry-thickening phase, as contrasted with a more quiescent filtration. Thus laboratory measurements via the Canadian Standard Freeness Test or the Schopper-Riegler Test, both of which essentially measure the rate of filtration, are not expected to provide close correlation with machine performance. Various approaches based on modifications of the Dynamic Drainage Jar (DDJ) have been reported (77) (78), but a considerable amount of interpretation is needed (79).Nevertheless, they are useful guides (108). "However, it is generally accepted that the best way to determine the effect of a polymeric additive on the moisture content of the sheet leaving the press section and entering the driers is by a mill trial" (72).

It has been pointed out that generating large flocs may provide rapid drainage in the first stage, but the relatively large diameter capillaries formed in the web allow more air to flow through during the high-vacuum dewatering stage, producing a wetter mat. Also, if the flocs are too large, they may interfere with formation. Floc size can be controlled by the selection of aid and of the addition point since there are many places in the paper machine which can break up large flocs (making a more uniform dispersion). Hydrodynamics of the system is an important consideration which must be tested on the machine itself.

Frequently, de facto drainage aids are present in an acid furnish, such as alum. Medium molecular weight C-PAMs are reported to give good overall drainage in bleached and Kraft furnishes (77). These drainage aids also enhance fines and filler retention, which probably helps in the high-vacuum dewatering stage. Since many of the same polymers function as both types of aids, the interrelationship is strong. Thus comments on the effect of microparticle

systems on dewatering will be deferred to the discussion of retention (80).

Retention

"Retention is a measure of the amount of solids in the web or sheet compared to the amount added at the wet end" (81). Concerned about the proliferation of ways in which retention is reported, a committee has been working to establish some uniformity (82). The following definitions are consistent with initial recommendations:

Overall or Comprehensive Retention. This is the ratio of the rate of production (tons/day) at the reel to the rate of addition of all solids. It is merely an overall material balance. Multiply by 100 to obtain percent.

Total Solids First Pass Retention. FPR(T) is defined by the expression:

$$\%FPR(T) = 100 \times [T(H) - T(W)]/T(H)$$

where $T(H)$ is headbox consistency and $T(W)$ is white water consistency.

Additive (A) First Pass Retention. FPR(A) is defined by the expression (81):

$$\%FPR(A) = 100 \times [A(H) - A(W)]/A(H)$$

where $A(H)$ and $A(W)$ are the concentrations of additive A in the furnish in the headbox and in the white water respectively.

Accepting the basic physicochemical processes of fiber-fiber, fiber-particle, and particle-particle binding, consider the simpler mechanical processes encountered in retention (81) (83).

1. "Sieving by the wire and buildup of a fibrous layer on the wire retains long fibers and larger fines."

2. "Entrapment of smaller fines and some filler within the lumens of the fibers and in the microfibrillar network on the surfaces of beaten fibers."

Both points one and two occur without the use of aids and, while effective for long fibers, they are limited for fines and fillers in the absence of aids. The function of retention aids is to facilitate the attachment of fines and fillers to the long fibers or to each other (flocculation) such that they remain in the web developed by Point 1. Generally, this flocculation occurs prior to the slice, and care must be exercised that problems in formation do not occur. This aspect can be controlled by choice of aid(s) and addition point. An additional feature with aids is the possibility of attachment of particles (filler or fines) to surfaces of the web during drainage. This is analogous to direct filtration in water treating technology (84).

Laboratory screening of retention aids for first pass retention with a given furnish via intelligent use of the DDJ can be a predictive tool (79). It does not, however, give information on paper quality nor can it reproduce the complicated hydrodynamics of a paper machine. Hence fine tuning can only be done during a mill trial on the specific machine for which aid(s) are to be used. The mill trial requires careful planning of specific objectives, handling of the aid(s), method of addition, method for establishing performance, machine conditions or operating changes, length of time needed to establish a steady state, etc.

Generalizations on usage of retention aids are summarized here on the basis of literature reports. There are complications to any generalizations, such as filler pretreatments to change surface properties, and occasional furnishes that exhibit unexpected behavior (85-88). Also, new developments in retention aid products, fillers, and sizes can cause changes (88).

Single Component

A recent review (72) states "...In most papermaking operations, especially in fine paper production, single polymer retention aids are the systems of choice. The simplicity and proven reliability of using the traditional high molecular weight cationic polyacrylamides in this system cannot be disputed." This quote applies particularly to acid papermaking, and the C-PAM generally has low cationicity (88). Many of these furnishes contain alum and cationic starch, added for other purposes, but which can also function as de facto aids and influence performance of the high molecular weight C-PAMs. Addition point usually is at Point A4 after the fan pump.

Polyethylene oxide is reported to be an effective single aid in furnishes which contain predominantly mechanical pulp (89). This furnish tends to have significant dissolved anionic species which can interfere with efficient use of C-PAM even though alum is used for pH adjustment. The PEO probably is added at Point A5, since PEO flocs are reported to be especially shear-sensitive.

In many alkaline papermaking furnishes, A-PAMs have a dominant position (88). Such furnishes generally have a low anionic loading (from chemical pulp) and moderate cationic loading (PCC without dispersant, cationic starch, alum). If the furnish contains mechanical fiber, secondary fiber with high anionic loading, or significant clay, other aids may be better. Among these are C-PAMs or a dual system.

Dosages for high molecular weight PAMs tend to fall in the range 0.1-0.6 kg/metric ton (0.2-1.2 lbs/ton) based on dry paper.

Multiple Components

Here we shall be concerned with retention systems which consist of deliberate additions of multiple aids (neglecting de facto aids).

Dual Polymer Systems

A recent review (72) comments "...In a dual polymer retention aid system, two synthetic polymers are mixed with the pulp sequentially to achieve better results than obtained with either polymer by itself. Usually a low molecular weight, highly charged cationic polymer is added to the papermaking furnish first and then, at a later stage, a high molecular weight, anionic polymer is added." "...dual polymer retention aids have found a place in paper and board manufacture..." This latter comment applies particularly to alkaline furnishes. Sometimes a high molecular weight C-PAM is used in place of the A-PAM (71) (90).

Mechanical pulps generally have significant concentrations of negatively charged dissolved and suspended materials known as "anionic trash" which place a high cationic demand on the retention aid system (85). The rationale for a dual system is to satisfy most or all of the extra cationic demand by a low molecular weight, very highly charged PEI or poly(DADM) or polyquat, and then use a high molecular weight A-PAM or C-PAM. Use of polyaluminum chloride (PAC) to treat anionic trash has been reported to exhibit an advantage in some furnishes (109). A rather different type of dual is PEO combined with a phenol-formaldehyde resin, which is also reported to be effective for furnishes of this type (89) (110-111).

Acid papermaking conditions, typically having pH<5, are less favorable to a cationic-anionic dual than for alkaline systems. "Dual synthetic systems can be effective where they can economically complement furnish components such as cationic starch and alum. The possibility of a relatively greater balance between anionic and cationic loadings in alkaline furnishes compared to acid furnishes is why these systems play a larger role than they do in acid papermaking. However, careful and thorough testing is required to determine if they actually are effective in the specific situations under consideration" (88).

Dosages for the low molecular weight cationic aid tend to fall in the range 0.1-0.8 kg/metric ton (0.2-1.6 lbs/ton), and for the high molecular weight PAM in the range 0.1-0.6 kg/metric ton (same range as for the single component system).

Microparticle Systems

In this type of multiple component system, a soluble cationic flocculant is used in conjunction with a solid (microparticle) anionic flocculant (91) (92). In this context, microparticle has the characteristics of a solid (fixed size and shape resistant to deformation) as contrasted with a soluble polymer random coil of comparable average size. The particle carries a negative surface charge when suspended in an aqueous medium, and counterions will be in solution. In terms of size, a range of 1-50 nm in diameter has been cited in patent claims with a preferred value in the range of 3-10 nm. If the particle is not spherical, such as a sodium montmorillonite platelet which is much larger, there may be a question as to the extent of disintegration of the stack of layers which make up the platelet. For present purposes, systems with bentonite will be included in this category. Examples of microparticle systems are cationic starch in conjunction with anionic colloidal silica (91) (112), C-PAM with sodium montmorillonite (93-95) (113), and cationic starch with anionic colloidal aluminum hydroxide (43) (107). There are also patent examples in which both A-PAM and anionic colloidal silica are used with cationic starch or C-PAM (96) (113)."

In practice, the cationic component usually is added first and allowed to interact with the furnish before addition of the anionic microparticles. For the cationic starch/colloidal silica system (COMPOSIL process), a typical dose of cationic starch is reported to be about 10-15 kg/metric ton (20-30 lbs/ton) based on dry paper, and of colloidal silica about 1-1.5 kg/metric ton (2-3 lbs/ton) (91). In an example of the C-PAM/bentonite system (HYDROCOL process), dosage is reported to be about 1-3 kg/metric ton (2-6 lbs/ton) of high molecular weight, medium cationicity C-PAM and about 2 kg/metric ton (4 lbs/ton) of a modified bentonite (94). In an example of the cationic starch, A-PAM, colloidal silica system for an alkaline fine paper furnish dosage was reported to be about 10 kg/metric ton (20 lbs/ton) of cationic starch, about 1 kg/metric ton (2 lbs/ton) of a high or very high molecular weight, medium anionicity A-PAM, and about 0.4 kg/metric ton (0.8 lb/ton) colloidal silica (96). The cationic starch/anionic aluminum hydroxide system (HYDROZIL process) can only be used in alkaline media. Colloidal aluminum hydroxide is formed *in situ* from alum and sodium hydroxide, with dosages reported in a range around 10 kg/metric ton (20 lbs/ton) cationic starch and 10 kg/metric ton (20 lbs/ton) alum for wheat-straw pulp (107)."

When compared to other dual retention aid combinations, it has been stated in the literature (92) that microparticle systems show the following characteristic features:

1. A strong positive dewatering effect is experienced both in the dryer section and in the press section.

2. The formed and dried sheet often exhibits a higher porosity.

Arguments for the use of the microparticle system hinge as much on improvements in drainage (allowing greater throughput) and formation as on retention.

But as with all aids, it is overall cost-performance which determines the value of additives and each situation has to be evaluated carefully. The increased cost of multiple component systems must be justified.

Saveall

The operation of the saveall is to provide a solids-liquid separation process (97). This may be via simple gravity sedimentation in a holding tank, sedimentation basin, or via a more complicated procedure. Those most commonly used are drum or disc-type filtration (98), or dissolved air flotation (DAF). Since paper fibers are close to water in specific gravity, the flotation process is used extensively in the paper industry (99). Excess white water from the paper machine flows into a collection tank from which it is pumped with air under pressure into an aeration tank. From the aeration tank, the white water with dissolved air flows to a flotation vat via inlet valves. "The inlet valves are designed to operate on the principle of a closed injector, actually creating a vacuum zone at the moment of air release, producing the very fine bubbles needed for optimum flotation" (99). It is a function of these fine bubbles to float the fibers and fillers to the surface of the flotation vat.

Flocculant addition usually is via a distribution header near the inlet valve. The flocculant function is to promote attachment of fibers-filler to the air bubbles, and normally high molecular weight N-PAMs or slightly ionic PAMs are used. In some cases, dual flocculant systems (cationic-anionic) have been used with the cationic component added near the inlet valve, and the anionic component added near the exit from the inlet chamber. Because of the variation in fiber-filler surface charge depending on additives to the stock, electrokinetic data are useful in selecting flocculants to test.

Filtration savealls tend to rely on pulp to build up the filter medium (sweetener), and so may not need a flocculant. If needed, addition is to the feed line to the saveall, or to the saveall headbox with agitation. Sedimentation units usually require conditioning with a flocculant, again added in the feed to the unit, with sufficient turbulence for good mixing.

Monitoring Mechanisms

On-line measurements can be an important aspect for control of paper machine operation and installation of equipment for this purpose is increasing (100). There appears to be a variation in actual use, however, which may be related to perceptions of reliability (101). In terms of measurements directly pertinent to retention aids, total retention and filler retention are available from standard, well-recognized procedures.

Ash

Ash is determined on-line by absorption of gamma radiation (from Fe-55) on passing through the paper sheet. Since ash has a higher atomic number than fiber, and the absorption coefficient increases rapidly with atomic number, this technique is specific to filler. Readings are calibrated in percentage ash content. Since feed data are constantly available, total filler retention is readily obtained (102).

Basis Weight

Beta-ray basis weight gauging has been generally accepted as a practical on-line tool. While the simplest form gives machine direction basis weight, a rapid off-line cross machine basis weight can be obtained from a beta-ray sheet weight profiler. Knowing the basis weight, along with the feed data, total retention is readily available (103).

White Water Flows

Standard flowmeters provide data on dewatering along with couch vacuum readings.

Experimental Techniques

Total First Pass Retention

A system for on-line measurement of consistency based on depolarization of a polarized light beam has been described. It has been installed on a fourdrinier-type machine whose furnish is made up of hardwood, softwood, refiner mechanical pulp, and coated and uncoated broke. Fillers include clay and titanium dioxide. One meter was installed on the headbox recirculation line to measure headbox consistency and another to measure white water silo consistency. A number of problems were encountered, but after working them out it is believed consistencies can be measured accurately enough to develop a first pass retention figure FPR(T), on-line (104).

Filler First Pass Retention

A continuous retention monitoring system providing separate measurements for first pass retention of total solids, FPR(T) and of filler, FPR(A), has been described. It is an optical system which depends on both attenuation and depolarization of a polarized beam (105).

Summary

Whereas drainage and retention are two distinct processes, they are interrelated in that the same aids affect both via basic flocculation phenomena. An aid selected for drainage may have less effect on retention than an aid selected for retention and vice versa, but it is a matter of degree. The performance of an aid is conditioned by the way it is handled and by the way it is used in a given mill as well as by its chemical composition. For acid papermaking, it appears that PEIs and C-PAMs are the main retention and drainage aids. Under alkaline conditions, A-PAMs have a dominant position for furnishes with low anionic loading. C-PAMs, PEO, and multiple component systems are more important for alkaline furnishes with high anionic loading. Microparticle systems appear to have some desirable features which need to be balanced against increased cost.

Reuse of water in the papermaking process has placed greater emphasis on efficient saveall operation. Sedimentation and flotation savealls are the main users of flocculants in this area, generally using high molecular weight, low ionicity products.

Acknowledgments

The invaluable assistance of Dr. D. S. Honig in preparation of this chapter is gratefully acknowledged.

Literature Citations

1. Halverson, F., and H. P. Panzer, "Flocculating Agents," In *Kirk-Othmer Encyclopedia of Chemical Technology*, 3rd edn., New York: Wiley, 1980, Vol. 10, pp. 489-523.

2. Everett, D. H., *Pure Appl. Chem.*, 31(3):579 (1972).

3. Farley, C. E., "Use of Polyelectrolytes to Optimize the Efficiency of Wet-End Chemical Additives," *TAPPI 1987 Papermaker's Conference Proceedings*, Atlanta: TAPPI PRESS, 1987, p. 295.

4. NBS, *J. Chem. Ed.*, 48(9):569 (1971).

5. Klass, C. P., A. J. Sharpe, and J. M. Urick, "Polyelectrolyte Retention Aids," *TAPPI C. A. Report No. 57*, Atlanta: TAPPI PRESS, 1975, Chap. 5.

6. Vorchheimer, N., In *Polyelectrolytes for Water and Wastewater Treatment*, W. L. K. Schwoyer ed., Boca Raton, Fla.: CRC Press, 1981, Chap. 1.

7. Molyneux, P., *Water-Soluble Synthetic Polymers: Properties and Behavior*, Boca Raton, Fla.: CRC Press, 1983, Vols. 1 and 2.

8. Schwartz, T., and J. Francois, *Makromol. Chem.*, 182(10):2757 (1981).

9. Halverson, F., and M. Botty, *Proc. Polym. Mat. Sci. Eng.*, 51:548 (1984), Fig. 2.

10. Halverson, F., J. E. Lancaster, and M. N. O'Connor, *Macromolecules*, 18(6):1139 (1985).

11. Truong, N. D., J. C. Galin, J. Francois, and Q. T. Pham, *Polymer* 27(3):467 (1986).

12. Lafuma, F., and G. Durand, *Polm. Bull.*, 21(3):315 (1989).

13. Panzer, H. P., and F. Halverson, "Blockiness in Hydrolyzed Polyacrylamide," Proc. Eng. Found. Conference Flocculation and Dewatering, Palm Coast, Fla., 1988.

14. Kurenkov, V. F., M. A. Nagel, and V. A. Myagchenkov, *Eur. Polym. J.*, 20(8):779 (1984).

15. Baron, J. J., and R. C. Montani, "Retention With Ultra-High Molecular Weight Polymer On A Twin Wire Machine," *TAPPI 1986 Papermaker's Conference Proceedings*, Atlanta: TAPPI PRESS, 1986, p. 79.

16. Yasuda, K., K. Okajima, and K. Kamide, *Polym. J.*, 20(12):1101 (1988).

17. Halverson, F., *Proceedings of Canadian Mineral Processor's Tenth Annual Meeting*, Ottawa, Canada, Jan. 1978, p. 404.

18. American Cyanamid Company, Work supported in part by NSF Grant No. CPE-11013 to Columbia University.

19. McCarthy, K. J., C. W. L. Burkhard, and D. P. Parazek, *J. App. Polm. Sci.*, 34 (3):1311 (1987).

20. St. Piepre, T., and M. Geckle, *J. Macromol. Sci. - Chem.*, A22(5-7):877 (1985).

21. Degen, H. J., and R. H. Lorz, "Specific Advantages of PEI Polymers in Highly Filled Paper Grades," *TAPPI l988 Papermaker's Conference Proceedings*, p. 363.

22. Dixon, J. K., G. L. M. Christopher, and D. J. Salley, *TAPPI Spec. Tech. Assoc. Publ.*, 31:412 (June, 1948).

23. *Jpn. Kokai*, 89 (04):207 (Jan. 9, 1989).

24. Gaspar, L. A., "Starch, Starch Derivatives, and Related Materials," *TAPPI C. A. Report No. 57*, Atlanta: TAPPI PRESS, 1975, Chap. 7.

25. BeMiller, J. N., "Gums: Industrial," In *Kirk-Othmer Encyclopedia of Chemical Technology*, 3rd edn., New York: Wiley, 1987, Vol. 7, pp. 589-613.

26. Coughlin, L. J., In *Handbook of Pulp and Paper Technology*, 2nd edn., K. W. Britt, ed., New York: VanNostrand Reinhold, 1970, pp. 631-641.

27. Sirois, R. F., and J. A. Janson, "Amphoteric Waxy Maize Starch: A New Dimension in Wet-End Starch Performance," *TAPPI 1959 Papermaker's Conference Proceedings*, p. 173.

28. Caldwell, C. G., W. Jarowenko, and I. D. Hodgkin, U. S. Patent 3,459,632 (Aug. 5, 1969).

29. Baes, C. F., Jr., and R. E. Messmer, *The Hydrolysis of Cations*, New York: Wiley Interscience, 1976.

30. Bottero, J. Y., J. E. Poirier, and F. Fiessinger, *Prog. Water Tech.*, 12:601 (1980).

31. Parthasarathy, N., and J. Buffle, *Water Res.*, 19(1):25 (1985).

32(a). Stol, R. J., A. K. VanHelden, and P. L. DeBruyn, *J. Coll. Interface Sci.*, 57 (1):115 (1976).

32(b). DeHek, H., A. K. VanHelden, and P. L. DeBruyn, *J. Coll. Interface Sci.*, 64 (1):72 (1978).

33. Akitt, J. W., J. M. Elders, and P. Letellier, *J. Chem. Soc., Faraday Trans.*, 83 (6):1725 (1987).

34. Arnson, T. R., *Tappi J.*, 65(3):125 (1982).

35. Strazdins, E. "Overview of Alum Chemistry in Papermaking," *TAPPI 1988 Papermaker's Conference Proceedings*, Atlanta: TAPPI PRESS, 1988, p. 382.

36. Reynolds, W. F., "Interaction of Alum and Papermaking Fibers," *TAPPI 1986 Papermaker's Conference Proceedings*, Atlanta: TAPPI PRESS, 1986, p.321.

37. Dentel, S. K., and J. M. Gossett, *J. Am. Water Works Assoc.*, 80(4):187 (1988).

38. Busler, W. R., "Aluminates," In *Kirk-Othmer Encyclopedia of Chemical Technology*, 3rd edn., New York: Wiley, 1978, Vol. 2, pp. 197-202.

39. Ban, S., *Nikkakyo Geppo*, 26 (1):27 (1973); *Jpn.Chem.Week*, (June 15, 1972).

40. Dempsey, B. A., H. Sheu, T. M. T. Ahmed, and J. Mentink, *J. Am .Water Works Assoc.*, 77(3):74 (1985).

41. Isao, M., H. Miyazaki, and H. Kawamura, (to Toyo Soda Mfg. Co. Ltd.) *Jpn. Kokai*, 77 99994 (Aug. 22, 1977); Kilby, B. J. L., (to Laporte Ind. Ltd.) Ger. Offen. 2,547,695 (Apr. 29, 1976).

42. Toa Gosei Chem. Ind. Ltd., 09.12.76-JA-147175; *Jpn. Kokai*, JP 88,275,795 (Nov. 14, 1988).

43. Lindstrom, T., "Aluminum-Based Microparticle Technology," *TAPPI 1989 Papermaker's Conference Proceedings*, Atlanta: TAPPI PRESS, 1989, p. 295.

44. Iler, R. K., *The Chemistry of Silica: Solubility, Polymerization, Colloid and Surface Properties, and Biochemistry*, New York: Wiley-Interscience, 1979, Chap. 3.

45. Stumm, W., H. Huper, and P. L. Champlin, *Env. Sci. Techn.*, 1(3):221 (1967).

46. Iler, R. K., *The Chemistry of Silica: Solubility, Polymerization, Colloid and Surface Properties, and Biochemistry*, New York: Wiley-Interscience, 1979, pp. 174-176.

47. Iler, R. K., *The Chemistry of Silica: Solubility, Polymerization, Colloid and Surface Properties, and Biochemistry*, New York: Wiley-Interscience, 1979, Chap. 4.

48. Moore, E. P., and G. Vurlicer, U. S. Patent 3,956,171 (May 11, 1976).

49. Chamberlain, R. J., "Polyelectrolyte Makeup and Handling," In *Polyelectrolytes for Water and Wastewater Treatment*, W. L. K. Schwoyer ed., Boca Raton, Fla.: CRC Press, 1981, Chap. 8.

50. McCollister, D. D., and co-workers, *Toxicol. Appl. Pharmacal.*, $\underline{7}$(5):639 (1965).

51. "Investigation of Selected Potential Environmental Contaminants: Acrylamides," EPA-560/2-76-008, TR 76-507 (PB 257704), EPA, Washington, District of Columbia.

52. Ramsden, D. K., and K. McKay, *Polym. Deg. Stability*, $\underline{15}$ (1):15 (1986).

53. Heitner, H., "Flocculants in the Presence of Fe^{+2} and Oxygen," Proc. Eng. Found. Conference Flocculation and Dewatering, Palm Coast, Fla., Jan. 1988.

54. Ottino, J. M., *Scientific American*, $\underline{260}$(1):56 (1989).

55. Chen, S. J., and A. R. MacDonald, *Chem. Eng.*, $\underline{80}$(7):105 (1973).

56. Calabrese, R. V., and C. M. Stoots, *Chem. Eng. Prog.*, $\underline{85}$(5):43 (1989).

57. Abdel-Alim, A. H., and A. E. Hamielec, *J. App. Polym. Sci.*, $\underline{17}$(12):3769 (1973).

58. Nagashiro, W., and T. Tsunoda, *J. App. Polym. Sci.*, $\underline{21}$(5):1149 (1977).

59. Long, J. T., *PCT Int. Appl.*, WO 89 01,456 (CA $\underline{111}$ 28137, 1989).

60. Kline, J. E., *Paper and Paperboard,* San Francisco: Miller-Freeman Publ., 1982, p. 136.

61. Strazdins, E., "Overview of Application of Electrokinetics in Papermaking," *TAPPI 1988 Papermaker's Conference Proceedings*, Atlanta: TAPPI PRESS, 1988, p. 379; "Applications of Electrokinetics to Papermaking," *TAPPI 1989 Retention and Drainage Short Course Notes*, Atlanta: TAPPI PRESS, 1989, p. 15.

62. Hubbe, M. A., "How Do Retention Aids Work?," *TAPPI 1988 Papermaker's Conference Proceedings*, Atlanta: TAPPI PRESS, 1988, p. 89.

63. Lindstrom, T., In *Fundamentals of Papermaking*, C. F. Baker and V. W. Punton, eds., London: Mechanical Engineering Publ., 1989, pp. 311-412.

64. Hubbe, M. A., *Tappi J.*, $\underline{69}$(8):116 (1986).

65. van de Ven, T. G. M., In *Fundamentals of Papermaking*, C. F. Baker and V. W. Punton, eds., London: Mechanical Engineering Publ., 1989, pp. 471-494.

66. Tam Doo, P. A., R. J. Kerekes, and R. H. Pelton, *J. Pulp Paper Sci.*, $\underline{10}$(4):J80 (1984).

67. Unbehend, J. E., *Tappi J.*, $\underline{59}$(10):74 (1976).

68. Pelton, R. H., *Tappi J.*, $\underline{67}$(9):116 (1984).

69. Marton, J. *Tappi J.*, $\underline{71}$(4):67 (1988).

70. Linke, W. F., *Tappi J.*, $\underline{51}$(11):59A (1968).

71. Honig, D. S., "Dual Polymers for Retention and Drainage: When and Why They Work," *TAPPI 1987 Advanced Topics Wet-End Chemistry Seminar Proceedings*, Atlanta: TAPPI PRESS, 1987, p. 1.

72. Allen, L. H., and I. Yaraskavitch, "Dual Polymer Retention Aids - A Review," *TAPPI 1989 Retention and Drainage Short Course Notes*, Atlanta: TAPPI PRESS, 1989, p. 59.

73. Alexander, S. D., and R. J. Dobbins, *Tappi J.*, $\underline{60}$(12):117 (1977).

74. Rahman, L., *Tappi J.*, $\underline{70}$(10):105 (1987).

75. Wegner, T. H., *Tappi J.*, $\underline{70}$(1):100 (1987).

76. Urick, J. M., and B. D. Fisher, *Tappi J.*, $\underline{59}$(10):78 (1976).

77. Scalfarotto, R. E., and R. F. Tarvin, *Tappi J.*, $\underline{67}$(4):80 (1984).

78. Wegner, T. H., A. M. Springer, and S. Chandrasekaran, *Tappi J.*, $\underline{67}$(4):124 (1984).

79. Honig, D. S., "Predictive Laboratory Retention Testing," *TAPPI 1989 Retention and Drainage Short Course Notes*, Atlanta: TAPPI PRESS, 1989, p. 29.

80. Sofia, S. C., U. S. Patent 4,795,531 (Jan. 3, 1989).

81. Stratton, R. A., "Introduction to Retention and Drainage Mechanisms," *TAPPI 1989 Retention and Drainage Short Course Notes*, Atlanta: TAPPI PRESS, 1989, p. l.

82. Kanitz, R. A., "Retention Definitions," *TAPPI 1989 Papermaker's Conference Proceedings*, Atlanta: TAPPI PRESS, 1989, p. 89.

83. Davison, R. W., *Tappi J.*, 66(1):69 (1983).

84. Tanaka, T. S., and M. Pirbazari, *J. Am. Water Works Assoc.*, 78(12):57 (1986).

85. Hagedorn, R. A., *Tappi J.*, 71(8):131 (1988).

86. Goodwin, L., *Tappi J.*, 72(8):109 (1989).

87. Griggs, W. H., *Tappi J.*, 71(4):77 (1988).

88. Honig, D. S., "Retention Aid Requirements for Alkaline Papermaking," *1989 Papermaker's Conference Proceedings*, Atlanta: TAPPI PRESS, p. 161.

89. Braun, D. B., and D. A. Ehms, *Tappi J.*, 67(9):110 (1984).

90. Wagberg, L., and T. Lindstrom, *Nordic Pulp Paper Res. J.*, 2(2):49 (1987).

91. Moberg, K., "Microparticles in Wet-End Chemistry," *TAPPI 1989 Retention and Drainage Short Course Notes*, Atlanta: TAPPI PRESS, 1989, p. 65.

92. Lindstrom, T., In *Fundamentals of Papermaking*, C. F. Baker and V. W. Punton, eds., London: Mechanical Engineering Publ., 1989, pp. 354-356.

93. Langley, J. G., and E. Litchfield, "Dewatering Aids for Paper Applications," *TAPPI 1986 Papermaker's Conference Proceedings*, Atlanta: TAPPI PRESS, 1986, p. 89.

94. Lowry, P. M., "Wet-End Balance Through the Use of Multi-Component Retention Systems," *TAPPI 1988 Papermaker's Conference Proceedings*, Atlanta: TAPPI PRESS, 1988, p. 231.

95. Ford, P. A., "The Application of a Dual Component," *TAPPI 1989 Retention and Drainage Short Course Notes*, Atlanta: TAPPI PRESS, 1988, p. 95.

96. Johnson, K. A., U. S. Patent 4,750,974 (Jan. 14, 1988).

97. Stevens, W. V., "White Water System Engineering," *TAPPI C. A. Report No. 57*, Atlanta: TAPPI PRESS, 1975, Chap. 3.

98. Byers, D., *Pulp and Paper*, 63(1):83 (1989).

99. Walzer, J. G., "Polyelectrolytes and Coagulants for the Flotation Process," In *Polyelectrolytes for Water and Wastewater Treatment*, W. L. K. Schwoyer ed., Boca Raton, Fla.: CRC Press, 1981, Chap. 5.

100(a). Lavigne, J. R., *An Introduction to Paper Industry Instrumentation*, San Francisco: Miller Freeman Publ., 1977, Chap. 19.

100(b). Lavigne, J. R., *Instrumentation Applications*, San Francisco: Miller Freeman Publ., 1979, Chap. 10, 12.

101. Shapiro, S. I., *Tappi J.*, 72(9):281 (1989).

102. Lavigne, J. R., *An Introduction to Paper Industry Instrumentation*, San Francisco: Miller Freeman Publ., 1977, pp. 284-285.

103. Lavigne, J. R., *An Introduction to Paper Industry Instrumentation*, San Francisco: Miller Freeman Publ., 1977, pp. 260-264.

104. Connolly, K. P., *Tappi J.*, 70(3): 89 (1987).

105. Kortelainen, H., J. Nokelainen, J. Huttunen, and K. Lehmikangas, *Tappi J.*, 72(8):113 (1989).

106. Trksak, R. M., "Aluminum Compounds as Cationic Donors in Alkaline Papermaking Systems," *TAPPI 1990 Papermaker's Conference Proceedings*, Atlanta: TAPPI PRESS, 1990, p. 229.

107. Wagberg, L., *Tappi J.*, 73(4):177(1991).

108. Allen, L. H., and I. M. Yaraskavitch, *Tappi J.*, 74(7):79 (1991).

109. Brower, P. H., *Tappi J.*, 74(1):170 (1991).

110. Carrard, J. P., and H. Pummer, U. S. Patent 4,070,236 (Jan. 24, 1978).

111. Stack, K. R., L. A. Dunn, and N. K. Roberts, *Appita*, <u>43</u>(2):125 (1990).

112. Sunden, O., U. S. Patent 4,388,150 (June 14, 1983).

113. Langley, J. G., and D. Holroyd, U. S. Patent 4,753,710 (June 28,1988).

114. BeMiller, J. N., "Gums: Industrial," In *Kirk-Othmer Encyclopedia of Chemical Technology*, 3rd edn., New York: Wiley, 1987, Vol. 7, pp. 597-598.

Chapter 9

Wet-strength Resins

by Robert E. Cates

Introduction

The hydrogen bonding that exists between cellulose fibers produces high levels of dry strength in most types of paper and board. However, these hydrogen bonds are sensitive to liquid water and water vapor, resulting in a severe loss of paper strength properties on wetting. Conventional papers saturated with water retain only 3-8 percent of their original dry strength, making them unsuitable for many uses involving water or exposure to high humidity.

Sizing agents are sometimes used to provide temporary wet strength by preventing water from soaking into the paper and contacting the fiber bonds. However, when there is prolonged contact with liquid water, the bonds eventually become wet and loss of strength results. For paper to retain strength properties upon long-term exposure to water or water vapor, it must be treated with commercial wet-strength resins. Although polymer chemists do not completely agree on the mechanism by which wet-strength resins function, wet-strength resins function via several possible mechanisms. Some examples follow:

1. Creating new fiber-resin-fiber bonds which are water-insoluble.

2. Surrounding existing bonds with a water-insoluble polymer network.

3. Functioning via a combination of both mechanisms.

With commercial wet-strength resins, it is possible to produce paper which retains up to about 40 percent of its tensile strength, 50 percent of its Mullen-bursting strength and 80 percent of its tearing strength when saturated with water. Paper that retains more than about 15 percent of its original tensile strength is considered to be wet-strength paper.

While wet-strength resin may occasionally be applied at the size press or in a separate tubbing operation, most wet-strength paper is produced by internal addition. The properties required in a wet-strength resin for internal addition are as follows:

1. It must be soluble in water.

2. It must be cationic to be retained on the pulp fibers.

Table 1. Typical Properties of Urea-Formaldehyde Wet-strength Resin

	25% Solids		35% Solids	
Total solids, weight %	25		35	
Ionic character	Cationic		Cationic	
Appearance	Amber liquid		Amber liquid	
Density at 77°F (25°C), lbs/gal	9.2		9.6	
Viscosity at 77°F (25°C), cps	15-30	(a)	35-60	(a)
pH	7.0-7.5		7.0-7.5	
Freezing point, °F (°C)	29° (-2°)		25° (-4°)	
Effect of freezing	Adverse	(b)	Adverse	(b)
Shelf life	3 months	(c)	3 months	(c)

Notes:
   (a) Freshly made. Viscosity increases with age, due to polymerization of the resin.

   (b) Resin stratifies and bottom portion will gel if not agitated immediately on thawing.

   (c) Guaranteed by most manufacturers if stored below 90°F (32°C). The resin slowly polymerizes and will eventually gel and become unusable.

3. It must cure to a water-insoluble state on drying of the paper, either by cross-linking with itself, cross-linking with cellulose, or both.

For more detailed information on the chemistry of wet-strength resins, the reader is referred to "Wet Strength in Paper and Paperboard," *TAPPI Monograph Series No. 29* ([1]) and *Pulp and Paper Chemistry and Chemical Technology* by James W. Casey ([2]).

Types of Wet-strength Resins

Types of wet-strength resins which are commercially available include:

Acid-curing
Urea-formaldehyde resin
Melamine-formaldehyde resin

Neutral or acid-curing
Polyamide-epichlorohydrin resin
Polyamine-epichlorohydrin resin
Acrylamide-glyoxal resin

Urea-Formaldehyde Wet-Strength Resin
Urea-Formaldehyde Resin Properties

This resin is a water-soluble urea-formaldehyde condensate containing a cationic modifier in the polymer chain. The cationic modifier is usually a polyfunctional amine such as diethylenetetriamine or triethylenetetramine. Urea-formaldehyde wet-strength resin is a syrupy liquid which is usually shipped at 25 or 35 percent solids. Typical properties are detailed in Table 1.

Urea-Formaldehyde Resin Storage Conditions

Urea-formaldehyde resin is shipped in bulk by tank car, tank truck, or in standard 55-gallon steel or fiber drums. The resin must be protected from freezing in transit and storage during the winter. It must also be protected from prolonged storage at temperatures above 90°F (32°C) during the summer. The resin may become completely unusable by gelation resulting from freezing, or by gelation resulting from storage at high temperatures.

Urea-Formaldehyde Resin Bulk Storage Tank Design

Urea-formaldehyde resin bulk storage tanks may be made of glass-reinforced plastic or Type 304 or Type 316 stainless steel. Mild steel tanks are unsuitable for storage of urea-formaldehyde resin due to corrosion above the liquid level and resulting discoloration of the resin. Paint-on inert resin linings have been successfully used to convert existing mild steel storage tanks, but are not recommended for new installations.

In cold climates, bulk storage tanks for urea-formaldehyde resin should be located indoors to prevent freezing. Inside storage tanks must be vented outside to comply with OSHA regulations.

In moderate climates, storage tanks for urea-formaldehyde resin are often located outdoors. The tank must be well-insulated to protect resin from possible freezing in the winter and to protect from excessively high storage temperatures in the summer. Glass-reinforced plastic tanks should have an extra layer of insulation built into the tank walls. Stainless steel tanks must have suitable external insulation.

Heating coils and heating systems for urea-formaldehyde resin are not recommended due to the danger of accidental overheating and gelation of the resin. If winter conditions are so severe that heating for the tank is indicated, the tank should be located indoors.

With prolonged use, bulk storage tanks for urea-formaldehyde resin will accumulate gelled material on the walls or on the bottom of the tank. Tanks should have a manhole and a large drain to enable occasional rinsing and cleaning.

Urea-Formaldehyde Resin Pumps

Due to ease of pumping and handling, urea-formaldehyde resin at either 25 or 35 percent solids is normally metered directly to the paper machine system without further dilution. Suitable corrosion-resistant pumps and pipelines must be used.

For pumping urea-formaldehyde resin, Type 304 or Type 316 stainless steel pumps are required. Centrifugal pumps are usually used. Positive displacement pumps such as gear, rotary, progressive cavity, diaphragm, or piston pumps are also satisfactory. In calculating the pump size, a maximum viscosity of 300 cps should be used for the resin, since it may become cold in storage and increase in viscosity, or resin may be stored past its intended shelf life and increase in viscosity. A centrifugal pump should have mechanical seals preventing sealing water from diluting the resin being pumped.

Piping for Urea-Formaldehyde Resin Bulk Systems

Piping for urea-formaldehyde resin may be plastic, plastic-lined steel, or Type 304 or Type 316 stainless steel. For outside storage tanks, all exposed piping should be well-insulated. This is to protect from potential freezing temperature in the winter and to protect from exposure to heat in the summer. Failure to properly insulate exposed piping can result in freezing and gelation in the pipelines in the winter, and overheating and gelation in the pipelines in the summer. It should be noted that resin piping which passes through hot areas of the paper mill may also be subject to overheating and gelation problems in either summer or winter.

For intermittent use of the resin, the system should be designed for cleaning out the piping after each use by flushing with fresh water. The system can also be designed with a recirculating loop back to the tank and resin can be kept circulating through the pipes at all times to prevent localized freezing or overheating.

Metering Urea-Formaldehyde Resin

Magnetic flowmeters are recommended for metering of urea-formaldehyde resin. The signal from the meter can be used to operate a control valve when a centrifugal pump is used, or to control the rpm of a positive pump. Any metering device which is viscosity-sensitive is not suitable, since the resin viscosity will change both with temperature and age of the resin. In particular, rotometers are not suitable for use with urea-formaldehyde resin.

Most urea-formaldehyde resin manufacturers recommend resin be diluted after metering to about 1 percent solids just before addition to the paper stock for more uniform distribution of the fibers. This can be accomplished by in-line addition of dilution water ahead of the point of addition to the stock, or by adding both the resin and dilution water to a mixing funnel at the point of addition.

The resin should be filtered before addition to the paper stock, preferably just ahead of the magnetic flowmeter and control valve. This is to remove insoluble gel particles which form in storage tanks and pipelines. Major suppliers of urea-formaldehyde resin can provide more detailed information on storage tanks, pumps, piping, and metering devices.

Using Urea-Formaldehyde Resin in Wet-strength Paper

Urea-formaldehyde resin belongs to the class known as acid-curing resins.

Chemically, this means that the condensation reaction which results in polymerization of the resin to the water-insoluble stage is initiated by hydrogen ions, which must be supplied by an acidic material. This material can be papermaker's alum or a mineral acid, such as sulfuric acid or hydrochloric acid. The headbox pH of the machine using urea-formaldehyde resin must be below pH 5 and preferably pH 4.0-4.5.

The actual cure of the resin is dependent on the pH of the paper itself. Depending upon the pH history of the pulp, the paper extract pH may not be the same as the paper machine headbox pH. For machines using urea-formaldehyde resin, it is recommended that actual pH of paper be determined regularly using TAPPI Method T435, "Hydrogen Ion Concentration (pH) of Paper Extracts."

Urea-formaldehyde resin is used in amounts of 0.5-3.0 percent, dry basis, depending on the wet-strength requirements of the paper grade. Typical usage is 0.75-1.5 percent.

Urea-formaldehyde resin is usually added at the fan pump of the paper machine. It may also be added to the thick stock, as at the stuff box. When rosin size is used, it is essential that the rosin size first be reacted with the alum before the addition of the urea-formaldehyde resin. Otherwise, the cationic urea-formaldehyde resin can react with the anionic rosin size resulting in the formation of a foamy complex and possible spots and deposits on the paper machine.

Major suppliers of urea-formaldehyde resin provide technical service to ensure optimum conditions for use of their product.

Testing Urea-Formaldehyde Wet-strength Paper

The polymerization of urea-formaldehyde resin to its final water-insoluble state is dependent upon the temperature history of the paper. On most paper and board machines, polymerization of the resin is not complete on the machine, and further polymerization occurs in the hot paper roll. In order to estimate the final level of wet strength that will exist in the paper, it is customary to give the reel samples an accelerated cure before testing for wet strength. Curing conditions of 5-15 minutes in a 220°F (104°C) oven are commonly used. The results of these artificial curing tests should be correlated with wet-strength tests run on samples of the paper or board after natural aging in the roll or after converting.

The most commonly used method of determining wet-strength properties is wet tensile. The wet Mullen test and wet tear test are also used. In order to measure the basic wet strength of paper, independently of temporary wet strength that might be produced by sizing, it is necessary that the specimen for wet testing be completely saturated with water. Wetting agents may be added to the water used for soaking, although this sometimes produces erroneous results. A vacuum-soaking method, in which the paper is immersed in water and exposed to several high-vacuum cycles, is preferred as a means of uniformly wetting sized paper or board. Absorbent papers such as tissue and towelling do not usually require special wetting procedures.

Repulping Urea-Formaldehyde Resin Broke

Due to its relatively slow rate of cure on the paper machine, fresh broke from wet-strength grades can usually be repulped without difficulty. Fully-cured broke, from rolls in storage or converting plant operations, may repulp with difficulty because of its higher levels of wet strength.

Fully-cured urea-formaldehyde resin is subject to hydrolysis under hot, acidic conditions. Fully-cured broke must usually be repulped by adding acid or alum to reduce the pH to 3-4 and heating the broke to 150°F (65°C) or higher, while subjecting to repulping conditions. The best equipment for repulping wet-strength broke is a high-attrition pulping device which is supplied by many manufacturers. To assure complete defibering of very high wet-strength grades, it may be necessary to follow the repulper with a broke jordan or deflaker.

Compatibility of Urea-Formaldehyde Resin

Urea-formaldehyde resin, concentrated at 25-35 percent, is highly cationic, and is also approaching the limit of its solubility in water, so that mixing with almost any other concentrated material in solution may result in precipitation of the resin. Concentrated resin will be instantly incompatible with almost all types of anionic materials, and mixing with concentrated solutions of salts, acids, or nonionic materials may also result in precipitation. Addition of any material which lowers the pH of the final mixture below pH 6 will result in polymerization, and gelation may occur within hours.

Dilute urea-formaldehyde resin is compatible with mildly anionic materials such starches and gum solutions, but incompatible with most strongly anionic materials. Lowering of the pH of a dilute urea-formaldehyde resin solution below pH 6 may also result in viscosity increase and possible gelation.

When used on the paper machine, care should be taken to keep the addition point of urea-formaldehyde resin well-separated from any highly anionic additives.

Table 2. Typical Properties of Melamine-Formaldehyde Wet-strength Resin

Form	Dry Powder	Acid Colloid
Concentration (solid)	100%	6-12%
Appearance	White	Tyndal blue solution
Density	0.4 Approx. (apparent)	1.05
pH	NA	1.6-2.0
Freezing point, °F (°C)	NA (NA)	32° (0°)
Effect of freezing	None	None
Shelf life	Indefinite	1-4 weeks at 75°F depending on solids

Urea-Formaldehyde Resin Spills

Since it is classified as a hazardous material, spills of concentrated or dilute urea-formaldehyde resin should be cleaned up as soon as possible. The resin is water-dilutable and it can be hosed to the sewer. Urea-formaldehyde resin is biodegradable and moderate amounts will not cause problems in waste treatment systems. Large spills could result in problems in biological treatment systems, due to the free formaldehyde content. Landfill disposal should be in accordance with all federal, state, and local regulations.

Safety Considerations with Urea-Formaldehyde Resin

Urea-formaldehyde resin is classified as hazardous by OSHA due to the presence of free formaldehyde in the resin. The fumes in the vapor space of a tank used for urea-formaldehyde resin are considered hazardous and appropriate OSHA tank entry procedures must be used.

Dilute concentrations of formaldehyde fumes from urea-formaldehyde resin are highly irritating to the eyes and nasal passages, and exposure should be avoided. Certain individuals are sensitive to formaldehyde, which acts as a skin irritant and sensitizer, and these individuals should avoid all contact with urea-formaldehyde resin.

The area in which the urea-formaldehyde resin is being metered to the stock should be well-ventilated, and contact with the resin should be avoided.

Government Regulation of Urea-Formaldehyde Resin

The manufacturer of urea-formaldehyde resin should certify that all ingredients are listed in the EPA TSCA inventory. Urea-formaldehyde resin is classified as hazardous by OSHA. An MSDS providing complete product details should be obtained from the supplier. Urea-formaldehyde resin claimed by the manufacturer to meet FDA requirements is suitable for use in the manufacture of food-packaging paper and paperboard.

Melamine Formaldehyde Wet-strength Resin[1]

Properties of Melamine-Formaldehyde Resin

This resin is available in two forms, a dry powder and a ready-to-use acid colloid form. The powder is dissolved in dilute acid solution and aged under specific conditions before using. In the acid colloid form the resin has a strong positive (cationic) charge. Typical properties for the two forms are detailed in Table 2.

Storage Conditions for Melamine-Formaldehyde Resin

The dry powder form is shipped in bulk or paper bags. It should be protected from moisture to avoid caking of the powder. For prolonged storage in excess of six months, high storage temperatures exceeding 100°F (38°C) should be avoided. Freezing is immaterial.

In the acid colloid form, the resin should be protected from temperatures greater than 80°F (27°C) and should not be stored longer than 10 days unless the resin is diluted to 6 percent concentration upon delivery. Ordinarily, the resin in this acid colloid form is shipped as soon as it is made from a plant located less than a day's transit from the paper mill. If the melamine-formaldehyde acid colloid is frozen, it can be thawed successfully provided localized high temperature is avoided. If the temperature is high enough, localized gelation occurs.

Melamine-Formaldehyde Acid Colloid Bulk Storage Tank Design

Bulk storage tanks may be made of glass-reinforced plastic, cypress wood, tile, or acid-resistant stainless steel. Inert resin linings have been used successfully for conversion of existing tanks, but are not recommended for new installations. Mild steel is not suitable due to corrosion below and above the liquid level.

In cold climates, bulk storage tanks should be protected from freezing to avoid the risks involved in thawing. Heating coils, heated surfaces, and live steam injection are not permitted.

Provisions should be made for occasional cleaning of the tanks to remove accumulated gelled or dried material. The tank should have a manhole and a large resin drain for occasional rinsing and cleaning. Before entering a tank, it must be thoroughly ventilated to remove formaldehyde fumes. National Institute of Safety and Health (NIOSH)-approved protective breathing apparatus should be used where formaldehyde levels are above the Permissible Exposure Limit (PEL).

Melamine-Formaldehyde Acid Colloid Pumps

Acid-resistant pumps suitable for handling the equivalent of dilute hydrochloric acid must be installed. Such pumps include gear, rotary, progressive cavity, diaphragm, piston and other positive-displacement pumps.

---

[1]Information in this section was provided by Walter F. Reynolds, American Cyanamid Company

Melamine-Formaldehyde Acid Colloid Piping

Acid-resistant pipe and fittings suitable for handling dilute HCl should be installed. Piping may be plastic, plastic-lined, mild steel, or Type 304L or Type 316L stainless steel.

Acid Colloid Preparation from Melamine-Formaldehyde Resin

Two formulas are shown in Table 3. One formula produces the regular acid colloid where the melamine-formaldehyde (M/F) ratio is 1:3. The second formula produces what is referred to as high efficiency acid colloid. The typical formula shown is based on the preparation of 100 gallons of solution in the proportion of one pound of trimethylol melamine (1:3 - M/F) per gallon (120 g/l). Suitable adjustments can be made to prepare batches of larger or smaller size.

Muriatic acid (HCl) at 20° Be' contains 31.45 percent hydrogen chloride. If acid of higher or lower HCl content is used, suitable adjustment should be made. The following simple directions for making up the resin solution should be followed closely.

1. HCl - If other than 20° Be' acid is used, suitable adjustment must be made. Caution should be exercised in handling HCl to avoid breathing fumes or contacting the eyes and skin.

2. Water should be measured carefully.

3. If necessary, batch size should be adjusted so that whole bags of resin are used. This avoids weighing the resin powder.

4. The sequence of adding of materials should be exactly as shown in Table 3. Under no circumstances should additional HCl be added after the resin has been added, since immediate coagulation is likely to occur.

5. Agitation should be thorough and should continue until the solution is complete (allow at least 30 minutes). Additional stirring is permissible.

6. Batch sequence. Fresh resin solution should not be prepared in a tank containing aged resin solution.

7. The fresh resin solution should stand for three hours (minimum time) before it is added to paper stock.

8. Other chemicals or resin solutions should not be mixed with the resin solution unless complete compatibility and stability have been previously ascertained .

The 12 percent resin acid colloid remains stable and usable for about one week if the temperature is 90°F (32°C) or less. Better stability is obtained at lower temperatures or at lower concentrations. If kept at higher solids or higher temperatures, the stability is reduced and gelling will occur in a shorter time. In practice, it has been found satisfactory to prepare the resin solution one day and use it for the next one or two days. In all cases, the resin solution should be allowed to age at 9-12 percent resin solids for at least three hours before diluting.

If the dry resin powder is lumpy (due to exposure to high humidity or water), it is advisable to add the resin through a screen of 1/4-inch mesh. Any friable lumps remaining on the screen should be crushed and allowed to pass through the screen. If the lumps are hard and glassy, they should not be used.

Using Melamine-Formaldehyde Acid Colloid in Wet-strength Paper

The aged resin solution is usually added to slush stock as close to the machine headbox as practical, while still allowing time for uniform distribution and adsorption of the resin. Passing resin-treated stocks through refiners and Jordans should be avoided. After metering, resin should be diluted to 1 percent solids before addition to the stock. Avoid overdilution and the use of white water.

Increasing the contact time between the resin and the fiber results in more complete resin adsorption and minimizes the slowing effect of the resin on the stock. Due to wide differences in mill equipment, the optimum point for resin addition may be at the beater, the stuff box (regulator box), the fan pump, the machine screen outlet, or the headbox.

High concentrations of salts, particularly sulfates, are detrimental if present at the point of resin addition, but salts such as aluminum sulfate (alum) are not objectionable if they are added after the addition of the resin. In some cases, where high sulfates (greater than 250 ppm) are present at the headbox or screen outlet, efficiency can be improved by adding the resin early and the required alum late in the stock system. Where rosin size is used, it is essential that the rosin size first be reacted with the of resin alum before adding the melamine resin solution. Otherwise, the cationic melamine resin acid colloid can react with the anionic rosin size, possibly resulting in spots and deposits on the paper machine.

The amount of resin required depends on the results desired and varies from a fraction of 1 percent (percent dry resin solids based on dry fiber weight) for such items as toweling to 5 percent for papers requiring an exceptionally high degree of wet strength.

Table 3. Formulas for Preparation of Melamine-Formaldehyde Resin Acid Colloid

	Regular Formula	High-Efficiency Formula
Water at: 70°F ± 20° (20°C ±10°)	88 gal (333 liter)	89 gal (337 liter)
Muriatic Acid (20° Be')	4 gal (15 liter) 39 lb (17.7 kg)	3.2 gal (12.1 liter) 31 lb (14.1 kg)
Formaldehyde (37%)	None	20 gal
Total, water acid formaldehyde	92 gal (348 liter)	112.2 gal (424 liter)
Trimethylol Melamine	100 lb (45.5 kg)	100 lb (45.5 kg)
Total, final volume	100 gal (379 liter)	120 gal (454 kg liter)
MF_3 per gallon	1.0	0.83
Approximate solids % MF_3	12%	10%

Note: A muriatic acid conversion is listed below

°B'e	Sp. Gr. 15.6°C	% HCl	Factor Based on 20°B'e Acid
18	1.1417	27.92	1.126
19	1.1508	29.65	1.06
20	1.1600	31.45	1.00
21	1.1694	33.31	0.944
22	1.1789	35.21	0.893

For most grades, 1-3 percent resin solids on fiber provide satisfactory results.

Effect of Sizing Agents, Alum, and pH

When properly used in conjunction with the melamine resin acid colloid, papermaking materials such as rosin size, starch, and alum do not appreciably affect the efficiency of the melamine wet-strength resin. However, in some cases, certain starches and anionic colloidal materials cause a net loss of resin by adsorbing the resin and then passing through the wire into the white water system.

The pH of the stock as it goes to the wire has a definite effect. As with all acid-curing resins undergoing a condensation reaction to attain an insoluble stage, the lower the pH, the more rapid the cure. Within the normal papermaking pH range of 4.5-6.0, the resin cures very rapidly and should give satisfactory wet strength after normal drying on the paper machine.

In many cases, the cure is not entirely complete at the reel, particularly if the headbox pH is in the 5-6 range. This may be checked by heating samples in an oven at about 250°F (121°C) for 10 minutes. If the wet strength improves after this treatment, wet strength in the roll may be expected to improve after a few days in storage.

Although high-sulfate concentration (>200 ppm) is detrimental if present before the resin is added, it is beneficial to have a small amount of sulfate in the stock, the optimum level being 25-100 ppm.

Repulping Melamine-Formaldehyde Resin Broke

Due to the relatively permanent wet-strength properties of melamine resin-treated papers, it is desirable to repulp wet-strength broke directly off the machine, since the wet strength is normally not fully developed. Repulping conditions of high temperature and low pH destroy the wet-strength property, Most papers can be defibered by heating at 170-190°F (77-88°C) in the presence of 2-3 percent alum (based on fiber). Excellent results are obtained by pressure cooking (rag cookers, digesters) for a few minutes, even without alum or other acidic materials, and then processing in a broke beater in the normal manner.

The best equipment for repulping melamine wet-strength broke is a high-attrition pulping device of the type supplied by a number of manufacturers. To assure complete defibering, it may be necessary to follow the repulper with a refiner.

Compatibility of Melamine Resin Acid Colloid

The resin acid colloid will not tolerate the addition of any chemicals including acid, alkali, salts or anionic substances. Only dilution water is permissible.

Spills of Melamine Resin Acid Colloid

All spills should be cleaned up immediately, keeping in mind the acidic nature (ca pH 2) and the presence of formaldehyde fumes. Hosing to the sewer dilutes the resin enough so that any reaction with other substances in the sewer should not ordinarily create a problem. Large spills could conceivably cause problems in biological treatment systems because of the free formaldehyde content, and should not be flushed to sewers.

Safety Considerations with Melamine-Formaldehyde Resin

Melamine-formaldehyde resin is classified as hazardous by OSHA due to the presence of free formaldehyde in the resin solution. The fumes in the vapor space of a tank used for melamine resin acid colloid are considered hazardous and appropriate OSHA tank entry procedures must be used.

Dilute concentrations of formaldehyde fumes are highly irritating to the eyes and nasal passages and exposure should be avoided. Individuals with a specific sensitivity to formaldehyde should avoid all contact.

Government Regulations of Melamine-Formaldehyde Resin

An MSDS should be obtained from the resin manufacturer and procedures outlined by the MSDS should be followed. Melamine-formaldehyde polymer has been granted a prior sanction under the terms of the U.S. Food Additives Amendment to the Federal Food, Drug and Cosmetic Act for use in the manufacture of paper and paperboard products used in food packaging. This prior sanction is noted in an order by the Commissioner of Food and Drugs in the U.S. Federal Register of March 1, 1960 and is incorporated in the regulations in 21 CFR 181.30.

Neutral or Acid-Curing Resins
Polyamide-Epichlorohydrin and Polyamine-Epichlorohydrin Resins

Since the properties and use of both these types of neutral or acid-curing resins are very similar, they will be handled as a single subject. For simplicity, they will be referred to as polyamide resin and polyamine resin.

Properties of Polyamide Resin and Polyamine Resin

These resins are cationic water-soluble condensates of an amino-polyamide or a polyamine, and epichlorohydrin. They are used in a manner similar to the urea-formaldehyde and melamine-formaldehyde resins previously described, except they do not require acidic conditions for further polymerization in the paper. Polyamide resin or polyamine resin may be used at pH 5-9.

Typical properties of polyamide-epichlorohydrin and polyamine-epichlorohydrin resins are detailed in Table 4.

Storage Conditions for Polyamide or Polyamine Resin

Polyamide or polyamine resin is shipped in bulk by tank car, tank truck, or in standard 55-gallon steel or fiber drums. The product must be protected from freezing in transit and storage. In warmer climates, the product must be protected from prolonged exposure to temperatures above 90°F (32°C). Temperatures above 110°F (43°C) cause quick gelation of the resin, while prolonged exposure to temperatures from 90-110°F (32-43°C) will cause hydrolysis of the resin and loss in efficiency.

Polyamide or Polyamine Resin Bulk Storage Tank Design

Glass-reinforced plastic tanks, which are suitable for the storage of acidic materials, may be used for the bulk storage of polyamide or polyamine resin. Type 304L or 316L stainless steel can also be used for bulk storage tanks if suitable welding rods and procedures to avoid preferential weld corrosion are used for fabrication. The resin manufacturer can provide details. Paint-on inert resin linings have been successfully used to convert existing mild steel storage tanks but are not recommended for new installations.

In cold climates, bulk storage tanks for polyamide or polyamine resin should be located indoors to prevent freezing. Indoor

**Table 4. Typical Properties of Polyamide-Epichlorohydrin
Polyamine-Epichlorohydrin Wet-strength Resin**

	12.5% Solids	25% Solids	35% Solids
Total solids, weight %	12.5	25	35
Ionic character	Cationic	Cationic	Cationic
Appearance	Amber liquid	Amber liquid	Amber liquid
Density at 77°F (25°C), lbs/gal	8.6	8.9	9.4
Viscosity at 77°F (25°C), cps	40-60 (a)	120-160 (a)	120-180 (a)
pH	4-5	2-4	2-4
Freezing point, °F (°C)	30° (-1°)	27° (-3°)	30° (-1°)
Effect of freezing	Adverse (b)	Adverse (b)	Adverse (b)
Shelf life	3 months (c)	3 months (c)	3 months (c)

Notes: (a)   Freshly made. Generally decreases with age. The resin may gel, however, if stored
above 110°F (43°C).

(b)   Resin stratifies and bottom portion may gel if not agitated immediately on thawing.

(c)   Guaranteed by most manufactures if stored below 90°F (32°C). Efficiency decreases
during prolonged storage.

storage tanks must be vented outside to comply with OSHA regulations.

In moderate climates, storage tanks for polyamide or polyamine resin are often located outdoors. The tank must be well-insulated to protect resin from possible freezing in the winter and to protect from excessive storage temperatures in the summer. Glass-reinforced plastic tanks should have an extra layer of insulation built into the tank walls, and stainless steel tanks must have suitable external insulation.

The installation of a heating system for outside storage tanks is not recommended due to the danger of accidental overheating of the tank contents resulting in either gelation or severe loss of efficiency. If winter conditions are so severe as to necessitate a tank heating system, tanks should be located indoors.

Suitably insulated tanks which receive regular shipments of polyamide or polyamine resin should not require a cooling system in order to keep the tank contents below 90°F (32°C). However, in extremely hot climates, or with low turnover of resin, the installation of cooling coils may be considered to maintain maximum resin efficiency.

On prolonged use, bulk storage tanks for polyamide or polyamine resin will accumulate gelled material on the walls or on the bottom of the tank. Tanks should have a manhole and a large drain to enable occasional rinsing and cleaning.

Polyamide or Polyamine Resin Pumps

Centrifugal pumps are usually used for handling of polyamide or polyamine resin. All wetted parts of the pump should be made from Type 304 or Type 316 stainless steel. The pump should have mechanical seals to prevent sealing water from contacting resin being pumped. Gear, rotary, progressive cavity, diaphragm, piston, and other positive-displacement pumps can also be used.

A maximum viscosity of 300 cps should be used in determining pump and pipe size. A system designed for this viscosity will handle either low or high-solids resin at the higher viscosities that exist during winter temperatures.

Piping may be plastic, plastic-lined mild steel, or Type 304L or Type 316L stainless steel. There have been problems with preferential corrosion of welds in stainless steel piping. Flanged fittings should be used, or the resin manufacturer's recommendation should be obtained concerning pipe welds.

Polyamide or Polyamine Resin Bulk System Piping

For storage tanks located outdoors, all exposed piping should be insulated to prevent possible freezing in winter or overheating of the resin in the summer. Freezing of the resin causes precipitation and gelation which may plug the pipeline. Heating of the resin above 110°F (43°C) may cause gelation and plug the pipeline. It should be noted that resin piping passing through hot areas of the paper mill may also be subject to overheating and gelation problems in either summer or winter.

For intermittent use of the resin, the system should be designed for cleaning out the piping after each use by flushing with fresh water. The system can also be designed with a recirculating loop back to the storage tank and resin can be kept circulating through the pipes at all times to prevent localized freezing or overheating.

Metering Polyamide or Polyamine Resin

Magnetic flowmeters are recommended for metering of polyamide or polyamine resin. The signal from the meter can be used to operate a control valve when a centrifugal pump is used, or to control the rpm of a positive pump. Any metering device which is viscosity-sensitive is not suitable since resin viscosity will change with both temperature and age of the resin. In particular, rotometers are not suitable for use with polyamide or polyamine resin.

Most polyamide or polyamine resin manufacturers recommend that resin be diluted after metering to about 1 percent solids before addition to the paper stock for more uniform distribution on the fibers. This can be accomplished by in-line addition of dilution water ahead of the point of addition to the stock or by adding both the resin and dilution water to a mixing funnel at the point of addition.

The polyamide or polyamine resin should be filtered before addition to the paper stock, preferably just ahead of the magnetic flowmeter and control valve. This is to remove insoluble gel particles which form in storage tanks and pipelines. Consult resin and filter equipment suppliers for recommendations of suitable filters.

Using Polyamide or Polyamine Resin in Wet-strength Paper

Polyamide and polyamine resins are known as neutral or acid-curing resins. Chemically, the resins are thermosetting, which means they will polymerize to a water-insoluble condition by the action of heat alone. Polyamide resin is more reactive than polyamine resin and is used in all types of pulps; polyamine resin is used primarily in unbleached grades.

Polyamide or polyamine resin can be used at a headbox pH of 5-9. A pH below 5 may adversely affect retention of the resin and rate of cure. The preferred pH is 6-8.

Polyamide or polyamine resin is used in amounts ranging from approximately 0.15-

1.5 percent, dry basis, depending on the wet-strength requirements of the paper grade. Typical usage is 0.25-0.75 percent.

Polyamide or polyamine resin is frequently added at the fan pump of the paper machine. It may also be added to the thick stock, at the stuff box, or at the machine chest, for longer contact time with the stock.

When rosin size is used, it is essential that the rosin size be reacted with the alum before addition of the resin. Otherwise, the cationic polyamide or polyamine resin may react with the anionic rosin size, resulting in the formation of a foamy complex and possible spots and deposits on the paper machine.

Major polyamide and polyamine resin suppliers provide technical service in establishing optimum use conditions for their product and optimum conditions for machine runnability.

Testing Polyamide or Polyamine Resin Wet-strength Paper

The polymerization of the polyamide or polyamine resin to its final water-insoluble state is dependent upon the temperature history of the paper. On most paper and board machines, polymerization of the resin is not complete on the machine and further polymerization occurs in the hot paper roll. In order to estimate the final level of wet strength that will exist in the paper, it is customary to give the reel samples an accelerated cure before testing for wet strength. Curing conditions of 2-10 minutes in a 220°F (104°C) oven are commonly used. The results of these artificial curing tests should be correlated with wet-strength tests run on samples of the paper or board after natural aging in the roll or after converting.

The most commonly used method of determining wet-strength properties is wet tensile. The wet Mullen test and wet tear test are also used. In order to measure the basic wet strength of the paper, independently of temporary wet strength that might be produced by sizing, it is necessary that the specimen for wet testing be completely saturated with water. Wetting agents may be added to the water used for soaking, although this sometimes produces erroneous results. A vacuum soaking method, in which the paper is immersed in water and exposed to several high-vacuum cycles is preferred as a means of uniformly wetting sized paper or board. Absorbent grades, such as tissue, toweling, or corrugating medium usually do not require special wetting procedures.

Repulping Polyamide or Polyamine Wet-strength Broke

Freshly made polyamide-resin or polyamine-resin-containing broke can usually be repulped without difficulty because the resin is not fully-cured on the machine. However, fully-cured broke from roll storage or converting plant operations may cause repulping problems.

Polyamide or polyamine resin is subject to hydrolysis under hot, alkaline conditions. If a high-attrition repulping device in good working order is available, the broke may be repulped satisfactorily by adding sodium hydroxide to pH 11 and heating to approximately 180°F (82°C) during repulping.

If repulping times under these conditions are excessively long, it may be necessary to use an oxidizing agent to break down the wet-strength bonds. Sodium or calcium hypochlorite may be used at 1.0-1.5 percent, based on the fiber. Maximum repulping rate is obtained at 120-150°F (49-65°C), as the lower temperatures help in maintaining chlorine residual. Since residual chlorine can damage the performance of wet-strength resins on the machine, broke repulped with hypochlorite should have the residual chlorine reduced to zero after repulping by addition of a reducing agent such as sodium sulfite.

To assure complete defibering of polyamide resin or polyamine resin broke, the repulper should always be followed by a broke jordan or a deflaker.

Compatibility of Polyamide or Polyamine Resin

The higher-solids polyamide or polyamine resins are approaching the limit of their solubility in water, so that mixing with almost any other concentrated material in solution may result in precipitation of the resin. In addition, both lower and higher solids polyamide or polyamine resin are highly cationic and are completely incompatible with almost all types of anionic materials. Addition of any material which raises the pH of the final mixture above the normal low pH of the resin may result in polymerization, with gelation possibly occurring within hours.

Dilute polyamide or polyamine resin is compatible with mildly anionic materials such as starches and gum solutions, but incompatible with most strongly anionic materials. Dilute polyamide or polyamine resin solutions are most stable below pH 5. A higher pH may result in a viscosity increase or gelation.

When used on the paper machine, care should be taken to keep the addition point of polyamide or polyamine resin well-separated from any highly anionic additives.

Spills of Polyamide or Polyamine Resin

Since it is classified as a hazardous material, spills of concentrated or dilute polyamide or polyamine resin should be cleaned up as soon as possible. The resin is water-dilutable and can be hosed to the sewer. Polyamide or polyamine resin is slowly biodegradable and moderate amounts will not cause problems in waste treatment systems. Large spills could result in problems in biological treatment systems because the resin can act as a bactericide. Landfill disposal should be in accordance with all federal, state, and local regulations.

Safety Considerations with Polyamide or Polyamine Resin

Polyamide resin and polyamine resin may be classified as hazardous by OSHA due to the presence of epichlorohydrin or its degradation products in the resin. These materials are toxic and may be skin, eye, and throat irritants.

The concentrations of epichlorohydrin or its degradation products in the vapor space of a tank used for polyamide or polyamine resin may be considered hazardous. Appropriate OSHA tank entry procedures must be used.

The area in which the polyamide or polyamine resin is being metered to the stock should be well-ventilated. Contact with the resin should be avoided.

Government Regulations of Polyamide or Polyamine Resin

The manufacturer of polyamide resin or polyamine resin should certify that all ingredients are listed in the EPA TSCA inventory. Polyamide or polyamine resin may be classified as hazardous under OSHA regulations. An MSDS providing complete product details should be obtained from the resin supplier. Polyamide resin or polyamine resin claimed by the manufacturer to meet FDA requirements is suitable for use in the manufacture of food packaging paper and paper board.

Acrylamide-Glyoxal Wet-strength Resin [2]

Properties of Acrylamide-Glyoxal Resin

This resin is a water-soluble condensate of polyacrylamide and glyoxal containing a cationic modifier in the polymer chain. The cationic group is introduced by copolymerization of acrylamide monomer and a basic vinyl monomer. The acrylamide-glyoxal wet-strength resin is a low-viscosity liquid shipped at 6-10 percent solids. Typical properties are detailed in Table 5.

Storage Conditions for Acrylamide-Glyoxal Resin

Acrylamide-glyoxal resin is shipped in bulk by tank car, tank truck, or in standard 55-gallon steel or fiber drums. If it is supplied as a 10-12 percent solution, it should be diluted upon receipt if storage beyond a few days is anticipated. After dilution, the resin is stable for extended periods of time depending upon concentration and storage temperature (see Table 5).

Once the resin has gelled, the resin is not water-dilutable and must be discarded. Several days prior to gelation, the resin will show signs of stringiness. At this stage, efficiency has not been impaired, but the resin should be further diluted if additional storage is anticipated.

Acrylamide-glyoxal resin solutions which have become frozen can often be reconstituted by mixing thoroughly after thawing. Although full efficiency can be restored, care must be taken to produce a clear homogenous solution.

---

[2]Information in this section was provided by Walter F. Reynolds, American Cyanamid Company.

Acrylamide-Glyoxal Resin Bulk Storage Tank Design

Acrylamide-glyoxal resin bulk storage tanks may be made of glass-reinforced plastic, stainless steel, PVC, rubber, wood or glass. Mild steel, brass, and aluminum must be avoided.

In cold climates, bulk storage tanks for acrylamide-glyoxal resin should be located indoors to prevent freezing. Inside storage tanks must be vented outside to comply with OSHA regulations.

The stored resin solution should be protected from heat. The tanks should be located in the coolest area practical. Thawing of frozen resin solution can result in partially gelled resin if localized overheating occurs.

Acrylamide-Glyoxal Resin Pumps

Because of its normally low viscosity, almost any type of pump is suitable, provided the pump is made of acid-resistant material. Centrifugal pumps should have mechanical seals to prevent sealing water from diluting the resin.

Acrylamide-Glyoxal Resin Bulk System Piping

Piping for acrylamide-glyoxal resin may be plastic, plastic-lined steel, or stainless steel. Copper alloy metal or aluminum is unsuitable. Where piping is exposed to below freezing temperature, it should be well-insulated to prevent freezing and solidification of the resin solution. Overheating with subsequent gelation problems should be avoided by keeping the piping away from heat.

If the resin is to be used intermittently, the system should be designed for cleaning out the piping after each use by flushing with fresh water. A recirculating loop back to the supply tank allows circulation of the resin to prevent freezing or gelation of resin in pipes.

Table 5.   Typical Properties of Acrylamide-Glyoxal Wet-strength Resin

Total solids, weight %	6-10				
Ionic character	Cationic				
Appearance	Colorless				
Density at 77°F (25°C), lbs/gal	8.7				
Viscosity of a 10% solution at 77°F (25°C), cps	25-60				
pH	3.0 - 3.5				
Freezing point, F° (C°)	30° (-1°)				
Effect of freezing	None				
		70°	80°	90°	
Shelf life (days to gel)	10% solids	16	8	4	days
	7.5% solids	42	24	12	days
	6.0% solids	105	65	34	days

Metering Acrylamide-Glyoxal Resin

Magnetic flowmeters and positive-displacement metering pumps are recommended. Viscosity-sensitive metering devices are not recommended.

Generally, the resin solution is metered at the resin concentration of the supply tank and then diluted to about 1 percent solids just before stock addition. In-line dilution just ahead of point of addition to the stock also works well. The ideal point of addition would be to thick stock that will not be returned to a chest such as in the accept side of the stuff or regulator box, where the resin can be well-mixed with the stock before dilution at the fan pump. Addition of resin to a stock chest or white water should be avoided. As a precaution against any foreign matter or insoluble gel particles, which can form in storage tanks and pipelines, an in-line filtering device should be installed before the resin flow is metered.

Using Acrylamide-Glyoxal Resin in Wet-strength Paper

Although acrylamide-glyoxal resin belongs to the class known as acid-curing resins, its high degree of reactivity (ease of cure) makes it feasible to produce wet strength at near neutral pH. Practically complete cure is attained at the reel at pH 5.0-5.5, making it quite unnecessary to operate at pH 4.0-4.5, as is required for urea-formaldehyde wet strength development. Actually, substantial development of wet strength is realized off the reel at pH 6.0-7.0 with full wet strength obtainable after a relatively short storage period.

Either acid or base is suitable for pH adjustment of the stock on the wire. Carbonates or bicarbonates are not recommended. An acid salt like sodium acid sulfate could be used in place of a straight acid.

Acrylamide-glyoxal resin is used in amounts ranging from 0.25-1.5 percent, dry basis, depending on the wet-strength requirement of the paper grade. Typical use is 0.5-1.0 percent.

Acrylamide-glyoxal wet-strength resin is especially suitable for the production of absorbent grades where wet strength is essential. A portion of the wet strength produced by this resin is considered temporary. For example, within 30 minutes after soaking in water, 20-30 percent of the initial wet strength is lost. This feature can have special value in disposable personal grades of paper.

Since as all paper machines are not alike, it is difficult to specify exact points of addition applicable to all machines. Experimentation on each machine will quickly establish the best point of addition for the resin. In general, cationic acrylamide-glyoxal resin needs sufficient time to become well-distributed on the fiber before the sheet is formed (10 seconds minimum contact time). In many cases, excellent performance has been obtained by adding the resin in-line to thick stock. In most towel and tissue applications, the resin-treated stock can be lightly refined, screened, and cleaned with no adverse effect. Heavy refining is not recommended. The resin should not be added to the wire pit or white water where it may be absorbed primarily on the fines. No appreciable adverse effect has been noted at stock temperatures as high as 130°F (54°C).

With the possible exception of very high groundwood-containing furnishes, cationic acrylamide-glyoxal resin is effective with all chemical pulps, both softwood and hardwood. As is true for all cationic resins, the presence of residual pulping or bleaching residues is detrimental. Provided secondary fiber is clean, the resin works very well and permits the increased use of inexpensive secondary fiber (waste paper).

According to manufacturer's literature, the resin does not change the retention or freeness characteristics of the stock to a significant degree.

Commonly used defoamers, softeners, slimicides, release agents, sticking agents, optical brighteners, dye fixatives, and dyes can be used, but should not be premixed with the resin or added to the stock in closely adjacent streams. Anionic surfactants, such as the napthalene sulfonate type used for pitch control, should be avoided. Talc is permitted. Internal strength additives such as starches (anionic or cationic), carboxymethyl cellulose (CMC) and acrylamide-based polymers are usually not needed. The acrylamide-glyoxal resin ordinarily provides sufficient dry-strength enhancement, making the use of any other strength additives superfluous.

Acrylamide-glyoxal resin is not sensitive to chlorine, but is sensitive to sulfites. Concentrations in the stock above 3 ppm should be avoided. Hexametaphosphate, as well as tetraphosphate, should be avoided. For maximum performance, the bicarbonate alkalinity of the sheet on the wire should not exceed 50 ppm (expressed as ppm $CaCO_3$). At pH 5.5 or lower, it is unlikely this alkalinity will be exceeded. Appreciable bicarbonate alkalinity in the sheet going to the dryers decomposes on heating and raises the pH above 7.0, thereby slowing the cure of the resin.

Compatibility of Acrylamide-Glyoxal Resin Solution

The addition of any substance must be avoided to prevent rendering the acrylamide-glyoxal resin unfit for use. Acid, base, or any acidic or basic salts are especially bad as are any polyphosphates or sulfite compounds. The co-use of other cationic polymers should be thoroughly checked. Anionic substances are not permitted. All formaldehyde-containing materials should be thoroughly flushed out before introducing acrylamide-glyoxal resin.

Yankee Dryer Release Control

Glue and release agents are effective with acrylamide-glyoxal wet-strength resin. Glue and the resin should be added at well-separated points in the stock system.

Chlorine has no adverse effect on the dryer coating. Phosphates, other than hexametaphosphate or tetraphosphates can be added internally to control dryer surface characteristics such as carbon streaks.

Acrylamide-glyoxal resin can be sprayed on the Yankee dryer as a very dilute solution to maintain a coating in non-wet-strength grades.

Felt Washing

Use only neutral or alkaline washes. Acid wash is not recommended, unless the felt has been previously washed with an alkaline solution.

Testing Acrylamide-Glyoxal Wet-strength Paper

Unlike other wet-strength resin-treated papers, the wet strength imparted by acrylamide-glyoxal resin consists of both temporary and permanent wet strength. For this reason, the paper should be wetted with water just before testing. This is done by applying water to the strip of paper already positioned in the tensile instrument just before the test. The wet tensile result is commonly referred to as initial wet strength. Ordinarily, acrylamide-glyoxal wet-strength resin is not employed in grades where the end use requirement calls for permanent wet strength (long-term soaking).

Under typical paper mill conditions, approximately 70-90 percent of the total wet strength imparted by acrylamide-glyoxal resin is developed on the machine. In many

instances, this rapid cure eliminates the necessity of aging the paper before shipping or converting. Should completely cured test values be desired, sample sheets may be artificially aged by oven heating for 3-5 minutes at 220°F (104°C) maximum. The resin cannot be artificially cured above the wet-strength value that will be obtained from natural aging. Oven temperatures above 220°F (104°C) can degrade the resin and result in a misleading lower test result.

Because acrylamide-glyoxal resin imparts more dry strength than other wet-strength resins, the wet-over-dry relationship obtained with acrylamide-glyoxal resin can be misleading. Best practice is to assess the dry and wet strength results individually.

Repulping Acrylamide-Glyoxal Resin Broke

Fully-cured acrylamide-glyoxal resin treated paper can be more rapidly repulped in conventional equipment than paper made with other wet-strength resins. Oxidizing agents or acid catalysts are not required.

Often no pH adjustment is necessary, and, in many cases, repulping can be accomplished at room temperature with regular process water. A shorter repulping cycle can be effected by either raising the water temperature to 125°F (52°C) or higher, or raising pH to 9-10 with any inexpensive alkali, or a combination of the two. Both bleached and unbleached papers can be repulped. If bleaching is desired, an alkaline hypochlorite can be used in place of the mild alkali.

Spills of Acrylamide-Glyoxal Resin

Resin spills are very slippery. Spilled material should be absorbed on an inert material and scooped up. The area should be thoroughly flushed with water and scrubbed to remove residue. The resin is water-dilutable and it can be hosed to the sewer. The resin is biodegradable and moderate amounts will not cause problems in waste treatment systems. Large spills flushed to the sewer could result in problems in biological treatment plants because of the free glyoxal content. Landfill disposal should be in accordance with all federal, state, and local regulations.

Safety Considerations with Acrylamide-Glyoxal Resin

The fumes in the vapor space of a tank used by acrylamide-glyoxal resin are considered hazardous and appropriate OSHA tank entry procedures must be used. Respiratory protection is generally not required during normal operations. To prevent skin contact, impervious gloves should be used. Eye or face contact should be avoided by using chemical splash-proof goggles. Recommendations in the supplier MSDS should be carefully heeded.

Government Regulations of Acrylamide-Glyoxal Resin

Manufacturers of acrylamide-glyoxal resin should certify that all ingredients are listed in the EPA TSCA inventory.

No Permissible Exposure Limit (PEL) has been established by OSHA. An MSDS providing complete product details should be obtained from the supplier.

The resin is approved for use in paper and paperboard intended for use in contact with foods under 21 CFR 176.170 of the Food Additive Regulations of the Food, Drug and Cosmetic Act. It may be used as a dry and wet-strength agent employed prior to the sheet forming operation in the manufacture of paper and paperboard in such an amount that the finished paper and paperboard will contain the additive at a level not in excess of 2 percent by weight of the dry fibers in the finished paper and paperboard.

Literature Citations

1. "Wet Strength in Paper and Paperboard," *TAPPI Monograph Series No. 29*, John P. Weidner, ed., New York: TAPPI, 1965.

2. Casey, James P., *Pulp and Paper Chemistry and Chemical Technology*, 3rd edn., New York: Wiley-Interscience, Vol. 3, Chap. 18, pp. 1609-1626.

Chapter 10

Latexes

by James J. Scobbo, Sr.

Introduction

Latexes can be classified into three major groups: styrene butadiene, vinyl acetate, and acrylic esters. Within each group, there are many variations demonstrating different properties in both the coating color and in the coated paper or board properties. Generally, almost all coating mills use latexes. Latexes are also used in saturation and beater addition.

General Description

Latexes can be used as binders in coatings, either as the total binder or in combination with natural binders such as starch, protein, or casein. Latexes are used in beater addition to give improved strength and specialized properties to the web. Latexes used in the area of web saturation produce discreet end use properties. Secondary benefits in using these materials in coatings are viscosity and flow modifications, increased strength, improved water resistance, increased holdout, improved wet rub, and improved gloss and smoothness.

Styrene butadiene latexes are emulsions composed of a varying ratio of styrene, butadiene, and carboxyl monomers such as acrylic acid, methacrylic acid, itaconic acid, and fumaric acid. These materials are reacted in water with anionic surfactants, which stabilize the system, and a catalyst, such as ammonium persulfate.

Vinyl acetate latexes are produced in a similar manner but use different monomers. These latexes can be vinyl acetate homopolymers or variations which could be acrylic esters, such as ethyl acrylate and butyl acrylate. Some versions may contain ethylene as the flexibilizing monomer. Acetates normally use a nonionic soap system.

Acrylic esters are also emulsions which contain primarily styrene and various combinations of acrylic esters, such as methyl, ethyl, butylacrylates,and methylacrylate. They can also contain carboxylic monomers, such as acrylic or methacrylic acids.

All three types of latexes are supplied in the form of a water-based emulsion. These are normally supplied at a solids content of 48-54 percent and a pH of 6-9. Most mills using latex as a binder require these materials to be approved under FDA paragraphs 176.170 and 176.180. The selection of a latex for use is predicated by the end

Figure 1. Bulk Storage System
Typical for tank truck or tank car shipments.

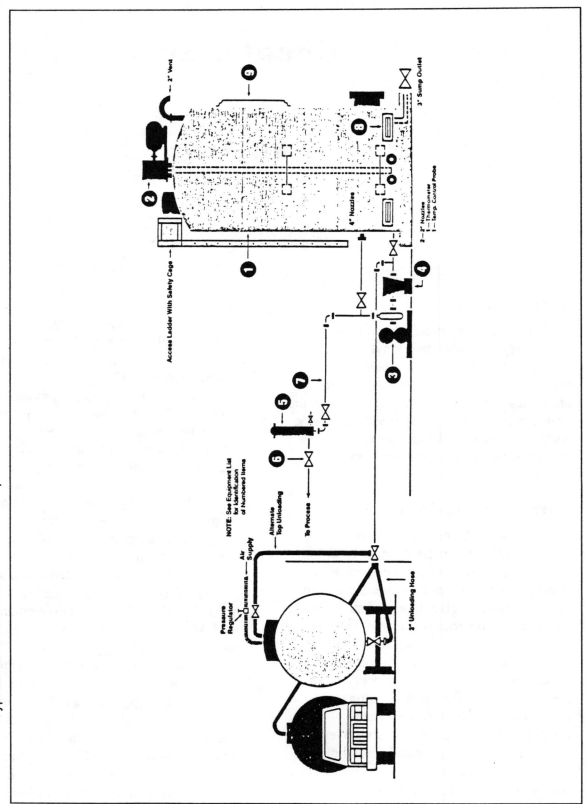

properties desired. Selecting the proper latex for a specific job is best accomplished by a joint program involving the user and his suppliers.

Storage and Handling

Latexes are available in steel drums or fiber packs, but most mills purchase them in bulk. This means they are normally shipped in tank trucks or rail cars.

A bulk storage system should be located as close as practical to the unloading area. Latexes are water-based and should be stored at temperatures of 65-85°F (18-29°C) and must be protected from freezing. Figure 1 details a typical bulk latex handling system featuring one storage tank. However, it is preferable to have more than one tank to facilitate cleaning. It is also recommended that vertical storage tanks be used to minimize the formation of skins on the latex surface.

Bulk Storage Equipment List
Storage Tank

Storage tanks may be constructed using a variety of materials. The most commonly used materials are stainless steel, lined carbon steel, and fiberglass. Figure 1 details a typical lined carbon steel tank installation.

A flat -bottom, dished-top head, vertical tank is most common. The tank should have a 1/4-inch wall thickness, fitted with a 20-inch diameter top manhole and a 20 to 24-inch diameter side-cleanout manhole agitation nozzle, 2-inch vent nozzle, 4-inch fill and discharge nozzles, 3-inch drain nozzle off bottom of tank, and two 2-inch nozzles for thermometer and temperature probes (outdoor tanks only). The tank is fitted with four baffles at 90° spacing each, 1/12 tank diameter in width, and tank straight-side dimension in length.

All interior tank surfaces, baffles, and nozzles should be coated with a 10-12ml epoxy-phenolic baked coating.

Tank Agitator

A low-speed (45 rpm) agitator with dual-turbine-type impellers of four-bladed turbine design with diameter equal to one-fourth times tank diameter is suggested. The use of stainless steel shafts and impellers is recommended.

Latex Pump

A low-speed (1150 or 1740 rpm) open impeller-type centrifugal pump with stainless steel wetted parts, and a 50-100 gpm capacity is suggested. Where adequate plant air is available, the air-operated double diaphragm pump may be used. It can be epoxy-coated and has the added advantage of eliminating shaft seal wear and leakage problems.

Latex may also be moved by pressure from the tank truck or rail car. Care must be taken not to exceed the pressure rating of the tank truck or rail car. Piston pumps are not recommended. The use of Moyno pumps for continuous pumping of latex is also not recommended.

Pump Strainer

A basket-type pot strainer with a removable basket drilled with 1/8-inch diameter holes is used.

Latex Filter

An in-line pressure filter with removable fabric sock of 60-80 mesh stainless steel mesh screen is suggested.

Latex Valves

Plug valves, ball valves, or diaphragm-type hand valves are suitable for latex service. Neoprene-reinforced diaphragms should be used in the diaphragm-type valves. Stainless steel or epoxy-lined valves should be used.

Pipe Fittings

Type 304 stainless steel, schedule 10 pipe is recommended for latex lines. Epoxy-lined carbon steel piping is satisfactory also. Plastic pipe, such as PVC, polyethylene, or fiberglass-reinforced piping also may be used. Pipe fittings should be the quick-connect type.

Tank Heaters

Heaters are mandatory wherever tanks are exposed to freezing temperatures. Single-embossed carbon steel Platecoil heaters are recommended. These are rolled to the outside tank diameter and bolted to tank walls with heat transfer cement sandwiched between. Hot water circulation is recommended.

Tank Insulation

For tanks located outdoors in areas of freezing temperatures, tank walls and top must be insulated with Fiberglass 6# density board or Foamglas block insulation at a 1 1/2-inch or 2-inch thickness. Two layers of reinforced glass fabric insulation seal are applied to cover this insulation.

Other Equipment

Latex Unloading Hoses. Two-inch minimum diameter, reinforced Neoprene hose or reinforced plastic hose is normally used.

Tank Measuring Device. Manual outage measurement from the top manhole is recommended using a windup steel tape with plumbob. Various differential pressure gauges or float level indicators are available.

Hose Connectors. Two-inch stainless steel hose quick connections are normally used.

Pipe Insulations-Outside Installations Only. Water Traced: Water tracing run on pipe with heat transfer cement between. Cover with 1-inch thick fiberglass with 15-pound asphalt roofing paper cover. Low-Pressure Steam Traced: Cover pipe with 1-inch fiberglass, run 1/4-inch copper tubing for steam tracing on insulation layer and cover with 1/2-inch thickness fiberglass with 15-pound asphalt paper cover. Steam tracing is not recommended as it may cause localized coagulation.

Latex Storage

Latex stored in a bulk storage tank should be maintained at a temperature of 65-80°F (18-27°C). Periodic mixing, 10 minutes every 24 hours is desirable to help minimize skinning of the latex and prevent possible stratification of material in the tank. During normal storage conditions, there should not be an extreme drop in pH. If, during prolonged storage, a situation occurs where the pH drops below specification, contact your supplier for special information. Latexes are normally stable, but it is recommended that these materials not be stored over six months.

Tank Cleaning

Latexes are treated with bactericide when manufactured and growth should not occur under normal conditions. However, in order to insure a trouble-free operation of the bulk handling system, it is recommended that this system be thoroughly cleaned and disinfected at least once a year. A description of the process follows:

The tank is emptied and all lines are blocked off. If tank surfaces have a heavy buildup, high-pressure jetting may be used. The interior is then steamed and air blown until the atmosphere is safe. The surfaces are then wiped off to remove remaining loose material, leaving tightly adhering material on the wall. Care should be taken not to damage the lining. The tank should then be disinfected. A solution of 50-75 ppm sodium hypochloride may be used for this purpose. This should be followed with a fresh water rinse.

Tank Truck and Tank Car Unloading

Receiving Shipments

Upon arrival of the tank truck or tank car shipment, the truck or car is spotted at the appropriate unloading spot. The top manhole should be opened and the shipment sampled according to the individual plant quality procedure.

Hookup for Tank Truck Unloading

After checking to be certain the tank truck outlet valve is closed, remove the cap over the outlet nozzle and attach the required length of clean, reinforced plastic or rubber unloading hose from the truck unloading nozzle to the transfer pipe leading to the suction of a latex pump. A two to three-inch I.D. hose should be used. NOTE: An on-board truck pump (gravity or compressed gas type) may be used in lieu of a plant latex pump.

Hookup for Tank Car Unloading

Remove the plug from the bottom outlet of the car after first making sure the foot valve and bottom gate valve are closed. Attach the required two or three-inch length of clean plastic or rubber-reinforced unloading hose from the car outlet nozzle to the suction of the plant latex pump or to the

inlet of the storage tank. NOTE: Normally in cold weather, the space between the foot valve and bottom gate valve will contain ethylene glycol antifreeze. The antifreeze should be drained out before attaching the unloading hoses.

Tank Truck and Car Unloading

Before beginning the unloading operation, the receiving tank should be gauged to determine if it will accept the full shipment.

Open the tank truck or tank car bottom valves and all other valves and disconnect the unloading hose from the empty truck or car. Bottom gate valves on the empty tank cars should be left open and the plugs left off for the return trip. Manhole covers and bottom foot valves, however, should be closed.

The unloading hose and lines should be rinsed with water after each use to reduce buildup due to air drying of a latex film on hose walls. Gauge the final tank measurement. Agitate the storage tank for about one hour to thoroughly blend the fresh latex with the older heel of latex in the tank.

Safety Precautions

Normally, latexes do not represent a health hazard. In the case of contact with skin or eyes, flush the area with water. If irritation occurs, seek medical attention. Latexes normally contain extremely low levels of residual monomers which should not pose a health problem, but inhalation of vapors and mists should be avoided. The water-based latex in its liquid form is not combustible, but the dried latex film is.

Small spills may be cleaned up by flushing with large amounts of water. Large spills should be dammed and the latex coagulated chemically and disposed of. Compliance with federal, state, and local regulations before disposing of waste materials in sewers and landfill is imperative.

Application and Use

Latexes are used primarily for binders in coating colors, size press coatings, beater additives, and saturation applications. In coating colors and size press coatings, the latex is normally added to the dispersed pigments and natural binder if it is being used. In beater addition, the latex is added to the machine chest or beater chest. In the saturation area, the latex may be formulated with other materials.

Latexes should be checked for total solids content, pH, and viscosity on a routine basis. If the solids content drops, it may be indicative of a leaky pump seal. A change in viscosity may indicate a drop in solids or a possible bacteria problem. Bacteria problems normally impart a hydrogen sulfide odor to the latex. The residue level in the latex should also be checked on a periodic basis to make sure the storage tank is not dirty.

Care should be taken that the level of latex in the storage tank is kept several feet above the outlet of the tank. If the tank level is drawn down too low, the cap of latex skins which form in the tank will be sucked into the pump and the latex system. If there is a filter on the outlet, it will plug. If no filter is present, this can cause a major problem in the coater operating area. Care should be taken to prevent mixing incompatible latexes. A high-pH latex should not be mixed with an alkali-swellable latex. This will result in extremely high viscosities and may plug the system. Latexes as a rule should not be mixed. If they will be mixed or there is a possibility of cross-contamination, a compatibility study should be conducted.

Sources of Additional Information

1. Latex supplier bulletins.

2. "Synthetic and Protein Adhesives for Paper Coating," *Tappi Monograph Series No. 22*, Atlanta, GA: TAPPI Press, 1961.

3. *Paper Coating Additives*, Atlanta, GA: TAPPI PRESS, 1982.

Chapter 11

Foam Control Agents

by J. R. Nelson

Introduction

Foam control agents are papermaking additives used in the manufacture of pulp, paper, and in paper coating formulations. Foam control agents are chemical formulations designed to prevent formation of entrapped air and surface foam (antifoams), or to destabilize existing entrapped air and surface foam (defoamers). Uncontrolled foam, or entrapped air, is a papermaking problem which can cause serious losses in productivity and profitability. Causes of foam in the papermaking system include surface-active materials from carrythrough of dissolved black liquor resins and soaps, complex carbohydrates, lignin, and other pumping chemicals.

Other causes of foam problems in the papermaking furnish are sizes and sulfonated additives. Acid pH paper machine operations have different additive foam problems than those in alkaline pH operations. Foam control agents are designed to reduce entrapped air, which causes sheet breaks or reduced machine speed, off specification paper due to holes and spots, or due to streaks caused by streams of minute air bubbles carrythrough. Foam control agents are available from many chemical suppliers. It is recommended that foam control agents be selected based on results using process (white water) sample tests, rather than on a random decision among available chemicals.

General Description of Additives

Purpose of Use

Foam control additives are designed to control entrapped air and surface foam in stock and in white water in order to provide trouble-free paper machine operation where foam problems are concerned. To be effective, foam control agents are added continuously, or in dual applications, one pump with a minimum dosage, with a second pump programmed to handle entrapped air surges. Application is normally to recirculating white water at the fan pump suction, but may be to other points depending on paper machine design considerations.

Description of Types Used

Foam control agents are typically made with a fluid carrier, a nonionic or anionic emulsifying agent, and a relatively insoluble additive which is actually the foam control chemical. Four basic types of foam control agents are available:

1. Oil-free or water-based.

2. Oil-water emulsions containing more than 50 percent water (may be called water-based).

3. Oil-water emulsions containing less than 50 percent water (often called water-extended).

4. Oil-based with no water.

Foam control agents usually are supplied as free-flowing fluids, although a few are available in brick or flake form. Easily pumpable liquids are preferred. Chemical and physical properties of foam control agents vary depending on the formulations. Users should carefully review product data sheets for specifics. In cases where pumping requirements are concerned, users are cautioned to be aware of temperature viscosity curves on fluids containing oil-water emulsions. One advantage of foam control agents as papermaking additives, is the improvement of production efficiency and increased capacity. In many cases, their use is necessary for the paper machine to operate. Foam control agents also help to achieve proper sheet qualities and specifications.

Disadvantages of foam control agents center around overuse. Excessive use may result in some oil combining with pitch or slime deposits. Improper use may also interfere with sizing efficiency. Also, carrythrough of some oil with stock in tissue machines may adversely affect protective coatings on Yankee cylinders. Prudent usage of foam control agents is important.

Regulatory Information

Most of the commercially available foam control agents are classified as non-hazardous by regulatory agencies such as the Department of Transportation (DoT) and by legislation such as the Toxic Substances Control Act. Users should always review foam control agent product literature and the supplier MSDS for more product details.

Criteria for Product Selection

Selecting foam control agents may be based on the following: history of chemical usage results on each paper machine, management decisions to use or avoid certain fluids and additives, stock temperature and pH conditions, fiber and other furnish materials, or other operating variables. It is preferable to prescreen candidates for foam control agents in on-site supplier's tests, although this is primarily a negative screen to avoid unnecessary work on machine tests or trials. The best time for selection of a foam control agent is when there is an entrapped air upsurge when white water tests can be made.

Storage and Handling Equipment
Shipping Containers

Foam control agents are shipped in 55-gallon drums, in 250 to 350-gallon semibulk containers (liqui-bins or tote-bins), in bulk containers by truck or rail, and in flexibins with bulk capacities.

Bulk Storage Equipment

Bulk storage equipment for foam control agents is available in capacities of 4,000-20,000 gallons and, typically, is sized to receive twice the volume of bulk delivery. Construction materials for storage tanks should be fiberglass, stainless steel, or polyethylene. The location of semibulk or bulk storage of foam control agents is important in that air temperature conditions should be in the range of 40-90°F (4-32°C). The distance from the storage tank to the application point should also be considered in planning an effective distribution system. Too great a distance could result in pump losses and inaccurate usages.

Feeding Equipment and Systems

Foam control chemicals are applied with metering pumps of many brands, often described as positive-displacement pumps. Pumps should have a filter with a 20 mesh screen to remove any contaminants. Pump suction lines should be sized to accommodate tho viscosity constraints of foam control agents. Use an inside diameter of one-half inch in the pump suction line to provide sufficient pumping flexibility with more viscous products, particularly under lower temperatures 40-60°F (4-16°C). Pump capacity should be twice a typical use rate. Flow measurement systems should be provided and monitored.

Construction Materials

For storage and handling, foam control agents should be held in and transferred through corrosion resistant tanks, lines, and pumps. Storage tanks should be equipped with suitable drains and ports for periodic cleanouts.

Safety and Precautions
Hazards and Handling Precautions

Hazards in handling foam control chemical agents are minimal. Formulations are surface active, and many contain oil. They may be slippery to handle. Users are referred to supplier's product literature for any specific hazards and recommended handling procedures. As with many chemical additives, workers should use rubber gloves and wear chemical splash goggles when handling foam control chemicals. Good ventilation is also very important.

First Aid

In case of accidental skin contact with foam control agents, wash thoroughly with water. For accidental eye contact, follow specific instructions from supplier product literature, MSDS, and product labels.

Handling Spills and Waste Disposal

Foam control chemicals are often used in waste treatment plants. Residuals normally dissipate before any effluent is discharged. Foam control agents based on oil may produce oil slicks from accidental spills into rivers and streams. Operator control to avoid spills is very important. Emulsifiers can be used in cases of bad spills to help break up oil slicks. In papermaking systems, foam control agents are consumed in the process due to their emulsifying actions. Chemically, they counteract the foam stabilizing materials that carry through with pulp and the pulping process. Residual waste materials in drums or bulk tanks can be used in waste treatment applications when such removal is needed. Waste from most foam control agents is considered nonhazardous.

Application and Use
Methods of Application

Foam control agents are available as concentrates or in direct application formulations. Concentrates must be diluted with water or other suitable fluid to achieve economic usage and control levels. Foam control agents are usually added on a continual basis with a control-metering pump. In coating formulations, or batch makeup of additives such as starches, foam control agents are also used on a ppm batch makeup procedure. Spray booms may be used to spread the application of foam control agents on vacuum washers or forming vat drums.

Treatment Rates

Foam control agents used in the papermaking process are added at a break-point dosage for the most economical use. A slight amount of foam in the wire pit is an indication that entrapped air is under control. Operator overuse must be avoided. Typically, usage levels vary by furnish and water conditions, but often are in the 0.5-2.0 pounds of foam control agent formulation per ton of paper. In cases of severe entrapped air problems, dosages can increase to 10 or more pounds per ton.

Points of Addition

Foam control agents are used in stock preparation, and in paper machine and white water applications. Typical addition points include screens, cleaners, pulp washers, decker vats, cylinder vats, stock lines, seal pits, trays, fan pump suction, and headbox showers.

Control Procedures

Applications of foam control agents should be controlled to avoid overuse and waste. Available control systems include basic electronic entrapped air liquid level controllers, and complex computerized systems that monitor foam control agent usage based on continuous measurements of white water entrapped air levels. White water or process foam demand systems are the most cost-effective ways to use foam control agents.

Evaluation of Effectiveness

Usage of foam control agents is determined by paper machine operating crews responsible for production capacity and quality. In the case of foam control agents, effectiveness means uninterrupted and continuous quality production. A lack of effectiveness means downtime and lost production

Potential Problems and Solutions

1. Foam control chemical agents application problems can occur in the delivery and pumping systems if inadequately designed. Careful attention must be given to the physical properties of each foam control agent, the transfer and application system, and pumping capacity in order to prevent such problems.

2. Attention to fluid foam control agent levels in storage tanks, and day tanks at the paper machines will ensure that pumping rates are consistent and uninterrupted. A flooded suction for foam control agent feed pumps may have a problem of "head" depending on the size of the tank and the vertical depth in the tank.

3. Adverse side affects may result from excessive use of foam control agents. This is particularly true in the case of surface deposits of fines and furnish materials.

Sources of Additional Information

1. Abrams, E. A., "Theory of Foam and Defoamers," 1958 Empire State TAPPI - Northern District Panel, 1958.

2. Nelson, J. R., "Let's Stop Overusing Defoamers," *Southern Pulp and Paper Manufacturer*, February 1975, p. 34.

3. Ellerby, R. W., "Foam - Some of its Causes, Consequences, and Control," *Pulp and Paper*, May 1974, pp. 59-61.

4. Twoomey, L. F., "Antifoam Application, Elements of an Effective Foam Control Program," *1986 TAPPI Papermaking Chemicals Process Aids Seminar Notes*, Atlanta, GA: TAPPI PRESS, 1986.

5. Nelson, J. R., "Trends in Regulated Control of Chemical Defoamers," *1990 TAPPI Papermaker's Conference Proceedings*, Atlanta, GA: TAPPI PRESS, 1990, p. 31-36.

Chapter 12

Creping and Release Aids

by J. R. Nelson

Introduction

Creping and release aids are papermaking additives used in the manufacture of tissue and towelling grades of paper produced on paper machines equipped with Yankee dryers. Creping aids are chemicals which coat the surface of the Yankee dryer and provide a functional continuous dryer film. This film increases the adhesion of the web to the dryer surface and provides more uniform drying. Creping aids are applied continuously either in the stock system or directly onto the surface of the Yankee dryer cylinder.

Release aids are chemicals which lubricate the cutting edge of creping blades, in order to maintain the integrity of the cutting edge in the creping function (see Figure 1). Inadequate or poor lubrication of the blade results in reduced blade life, more downtime for blade changes, and lesser crepe quality. Release aids are applied continuously either in the stock or white water system or directly onto the surface of the Yankee dryer cylinder.

The controlled and proper use of creping and release aids results in improved sheet crepe quality, reduced down time, and increased production. Creping aids can also provide a protective film to the Yankee cylinder surface reducing abrasive wear and corrosion. With the high capital cost of Yankee dryer cylinders, increased dryer life is of great interest everywhere.

General Description of Additives
Purpose of Use

Creping aids are additives which form a continuous film on Yankee dryer cylinders thereby increasing the adhesion of the web to the surface of the Yankee dryer cylinder. Release aids are additives which lubricate the cutting edge of the creping blade.

Types of Creping and Release Aids

Creping aids may be synthetic or natural polymers of various types and different molecular weights. Glues and wet-strength resins have been used. Creping aids are supplied in fluid solution or emulsion form. Release aids typically are oil-water emulsions and may contain special additives for high-temperature lubrication and metal protection. They are supplied in emulsion form. Newer

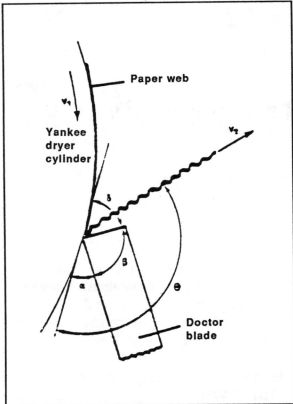

Figure 1. Creping Doctor Blade Geometry

synthetic chemical lubricants are available. Emulsions are designed to be stable in application, but to split at the high temperature of the creping blade-dryer interface >400°F (>204°C) thereby giving desired lubricity to the cutting edge of the blade.

Regulatory Information

Creping and release aids may be retained in the sheet. All components or ingredients must therefore be non-toxic. Components cannot be primary irritants or skin sensitizers. As a rule, medical clearance departments of tissue manufacturing companies issue

confidential assessments and use authorizations to production mills regarding the use of creping and release aids. Some tissue companies also require these additives to meet the regulations of the 1958 Food Additives Amendment on Indirect Additives, particularly when the production grades are towelling or lightweight food wrap.

Product Selection Criteria

Creping and release aids are typically machine specific and application programs are optimized by machine conditions such as pH, furnish, speed, internal dryer temperature ratings, and metal surface conditions where web and sheet contact takes place. It is strongly recommended that knowledgeable chemical suppliers, or specialized consultants, be employed to set up and optimize these papermaking additives uses with the paper machine crews and supervisors.

Storage and Handling Equipment

Shipping Containers

Creping and release aids are shipped bulk in 6,000-20,000 gallon (23-76 kiloliter) quantities, semibulk in liqui-bins of 250-350 gallons (1.0-1.3 kiloliters), or in 55-gallon (210 liter) drums. Proper labeling is required for interstate shipments.

Bulk Storage Equipment

Creping and release aids may be stored in bulk quantities provided such storage includes agitation or recirculation, and provision for annual clean outs. Creping aids normally are water solutions, so storage temperatures must be greater than 35°F (2°C). Also, creping aids being high molecular weight polymers, will show temperature X viscosity relationships. Lower temperatures of less than 50°F (10°C), may cause significant increases in viscosity, which can affect pumping and use rates. Adjustments may be required in application delivery systems.

stock or white water, they are typically pumped directly to the fan pump suction. Usage levels, may be five to ten times those of spray applications.

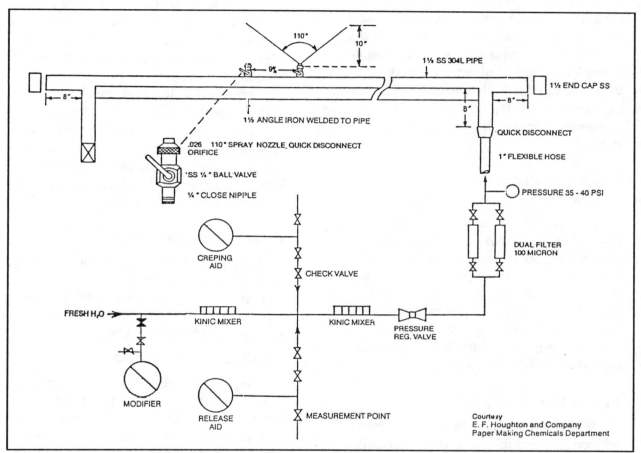

Figure 2. Spray Application System for Creping and Release Aids
Courtesy of E.F. Houghton and Company, Paper Making Chemicals Department

Feeding Equipment and Systems

The design of application systems for creping and release aids is important particularly where these aids are applied directly to the surface of the dryer cylinder. A schematic diagram is shown in Figure 2 (courtesy of E. F. Houghton & Co). Key points are mixers, pressure regulator valves, filters, and quick-disconnect spray nozzles. When creping and release aids are applied to

Construction Materials

For storage and handling, creping and release aids should be held in and transferred

through corrosion-resistant tanks, lines, and pumps. Lined mild steel, stainless steel, or fiberglass storage tanks are needed for bulk storage.

Safety and Precautions
Hazards

Creping aids are water solutions of natural or synthetic polymers and generally not considered hazardous. However, care should be taken to carefully read and review cautions listed in supplier product literature. Oil-water emulsion release aids may exhibit characteristic mildly objectionable odors when subjected to high temperatures in areas lacking good ventilation, such as the area immediately beneath the creping blade station.

Handling Precautions

In the use of creping and release aids, supplier handling precautions should be covered in detail with shift workers and supervisors. Such training meetings can be held well in advance of planned applications. The importance of optimum creping and release aid usage levels is stressed here. Higher pump rates or use levels do not always give better results. More is sometimes less where results are concerned.

Protective Equipment

Where workers may have direct contact with creping and release aids, eye protection goggles are recommended. Good ventilation is also very important.

First Aid

For accidental skin contact with creping and release aids, wash with large quantities of fresh water. Seek medical aid. For accidental eye contact, follow specific first-aid directions from the supplier literature and product label. Seek medical aid.

Handling Spills

Accidental spills of creping aids in working areas should not be hosed down with fresh water. Such efforts will result in extremely slippery floors and surfaces. Use of absorbents is recommended. Spills resulting in discharge to waste treatment plants may cause temporary upset (flocculation) in operations, but few serious problems. Spills to rivers or streams normally do not result in problems. Always refer to supplier product literature for specific handling information.

Accidental spills of release aids in working areas to waste treatment plants, to rivers, or streams should be handled as any oil spill. Use of absorbents is recommended on any exposed surface used by workers. Refer to supplier product literature for specific handling information.

Waste Materials Disposal

Creping and release aids, when used as directed, present no problem with disposal. The products are consumed in the process.

Application and Use
Application Methods

Creping and release aids historically have been used with wet-end addition and normally added directly to the fan pump suction for the best mixing possible with dilute stock. Carried through with the fiber furnish, these additives are transferred to the surface of the dryer cylinder as the web makes contact with and is drawn to the cylinder surface. The high surface temperature of the cylinder serves as a catalyst to the chemical reaction with the cylinder surface. To accomplish the intended technical effect in a cost-effective manner, high dosages of the creping and release aids are required.

Modern applications are more cost-effective, with controlled applications to the surface of the dryer cylinder by spray boom-bar systems installed beneath the dryer cylinder. Figure 2 illustrates how this is done.

Treatment Rates

Creping aids are typically used in dosage rates of 2-10 pounds per ton. Release aids are used in the range of 1-10 pounds per ton.

Points of Addition

The preferred addition points are directly to the cylinder surface as show in Figure 2. Alternatively, these aids can be applied to the fan pump suction, but with higher use levels.

Control Procedures

Addition to the dryer surface permits operators to have immediate control of additive usage and dosage changes depending on crepe performance and draw control.

Evaluation of Effectiveness

The evaluation of creping and release aids performance is determined by machine operation and by results achieved. Crepe control is measured by the physical costs of creped tissue or towelling produced and other qualities as specified. Longer-term effectiveness of creping and release aids on the operation of the Yankee dryer cylinder life, the creping blade performance, paper machine downtime and production gains are all determined by production reports and reviews.

Potential Problems and Solutions

1. Solutions of natural polymer creping aids may show microbiological deterioration and malodors. Good maintenance and clean outs of storage tanks are essential.

2. Excessive creping aid application dosages may give too much tack and

can wrap the sheet on the dryer. Note that controlled usages will provide more uniform web drying and improved crepe control.

3. One disadvantage with release aids is that excessive use on the machine may cause the sheet to flare off the blade and interfere with the creping aid film integrity. Controlled usages are important.

Sources of Additional Information

1. Nelson, J. R., "Creping and Release Aids," *1986 TAPPI Papermaking Processing Aids Seminar Notes*, Atlanta: TAPPI PRESS, 1986.

2. Valmet-KMW AB, "Pilot Tissue Machine," from an internal publication.

3. Uddeholm Strip Steel AB, "Creping Doctor Blades," from an internal publication.

4. Code of Federal Regulations, 21 CFR 170-191, (1989).

Chapter 13

Softening and Debonding Agents

*by John S. Conte
and
Gregory W. Bender*

Introduction

Pulp and paper producers have found it necessary to incorporate special materials into the wet end of their fibrous mass, in order to impart such properties as bulk, reduced mullen, and softness to the finished sheet. These properties are important in the production of sanitary papers and fluff or debonded pulp. Cationic surfactants will impart these unique properties to a wide range of papers through proper use.

The ability of the cationic surfactant to impart these properties is due to the fatty chain which is combined with the nitrogen group to obtain a cationic molecule. The cationic portion of the molecule, ammonium salt, carries a positive charge which allows it to attach itself to the surface of the weekly anionic cellulose fiber. This process is known as substantivity.

The fatty chain forms a thin film lubricant on the fiber surface which prevents extensive inter-fiber bonding and imparts a soft feel to the fiber. The attachment of the cationic material to the fiber occurs in dilute aqueous suspensions and is exhausted from the water as it is absorbed on the negatively charged surface of the cellulose fiber.

Softness

For many years, manufacturers of sanitary papers (facial tissues, toilet tissues, towels) used virgin pulp as a raw material. This offered excellent control over the fibers used in the furnished blend. However, with the increase in the price of this pulp, manufacturers recognize the need to reduce fiber cost. In order to do this, changes are made in the furnish. Some paper mills use recycled fibers as a cheaper alternative to the preferred virgin fibers. By doing this, softness is adversely affected.

The use of secondary fibers also leads to a variation in quality of the furnish obtained from lot to lot. This requires the manufacturer to vary the refining and the amount of dry or wet-strength resins needed to meet specifications. The increased use of these additives and refining also has an adverse effect on softness.

The use of cationic surfactants in the manufacture of sanitary papers made from secondary fiber yields a product which has a soft hand feel. This is accomplished through the lubricating nature of the substantive softening molecule, less extensive inter-fiber bonding leading to greater bulk, and the plasticizing effect these additives have on various resin systems.

Fluff or Debonded Pulp

Fluff or debonded pulp has the inherent characteristic of bulk, softness, high absorbency, and resiliency. Resiliency often depends on the length, diameter, and stiffness of the fiber. Long, stiff fibers will provide greater bulk and resiliency due to their distance to compaction. The inter-fiber voids formed in fluff or debonded pulp determines to a large extent the absorbency of the pulp. Large void areas lead to higher absorbency since it is these void areas that hold the moisture.

To produce fluff or debonded pulps of acceptable quality, it is important to optimize sheet bulk and minimize inter-fiber bonding. This is accomplished by such methods as running open or at reduced pressure on the wet presses, skipping wet presses, optimizing the draw, and using a chemical debonding agent. Generally, these conditions lead to slower production rates than for conventional pulp.

The mechanical fluffing of the pulp requires a pulp that will debond to a specific degree with the minimum power input and the least amount of mechanical fiber damage. Such a pulp must have the proper bulk and degree of inter-fiber bonding. A hard sheet will increase the power usage and increase fiber damage. Too soft a sheet will lead to pull-out of large pieces of pulp, causing poor fluffing.

Cationic surface-active agents are used in the manufacture of fluffed debonded pulp to improve bulk, reduce inter-fiber bonding,

and to soften the fibers. The lubricating effect of these agents prevents the formation of extensive inter-fiber bonding and increases the bulk of the sheet during formation on the machine. In the fluffing operation, the product improves the debonding characteristics of the sheet. This results in lower power requirements and less fiber damage. Reduced fiber damage produces a fluffed pulp with better bulk and resiliency.

General Description of Additives
Purpose of Use

The primary purpose for adding cationic materials in the wet end of the fiber mass is to impart such properties as softness, reduced mullen, and bulk to the finished sheet. These materials have also been found to improve the release of the sheet on the Yankee dryer.

In the application for the manufacture of fluff or debonded pulp, it is the cationic molecules added to the wet end that are used primarily to reduce the inter-fiber bonding of the sheet. This property is normally associated with a significant reduction in mullen strength. Since this pulp, in its converting operation, requires a significant amount of energy to convert it to final product, the use of fluff or debonded pulp reduces overall energy costs of conversion.

Description of Types Used

Although there are a number of cationic-type surfactants, there are four chemical types which are usually used to soften the pulp sheet or to produce a fluff or debonded pulp (see Figures 1-4). All materials are quaternary ammonium compounds, such as a nitrogen ion attached by covalent bonds to four organic groups. An anion, usually chloride, methyl, or ethyl sulfate, is associated with the positive ion of the quaternary nitrogen.

Supplied Forms of Agents

These cationic surface-active agents can be supplied as liquids, pastes, powders, solutions in water and alcohol, or solutions in water alone. They can be prepared as a solution in isopropanol or solvent-free.

Liquid is the preferred form due to ease of handling. The paste can be converted to liquid form with slight warming. Powders require much higher temperatures to achieve a liquid state.

Chemical and Physical Properties

Chemical and physical characteristics of these materials vary and are described in Table 1.

These chemical and physical characteristics primarily describe those products which are on the market today and being used by the industry for both debonding and softening.

Advantages and Disadvantages

Although there are many general types of quaternary ammonium materials which can be used to improve softness and debond pulp, there are advantages and disadvantages to each type.

Within the framework mentioned earlier, some advantages for using these types of

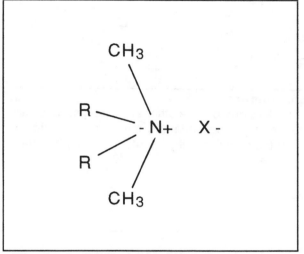

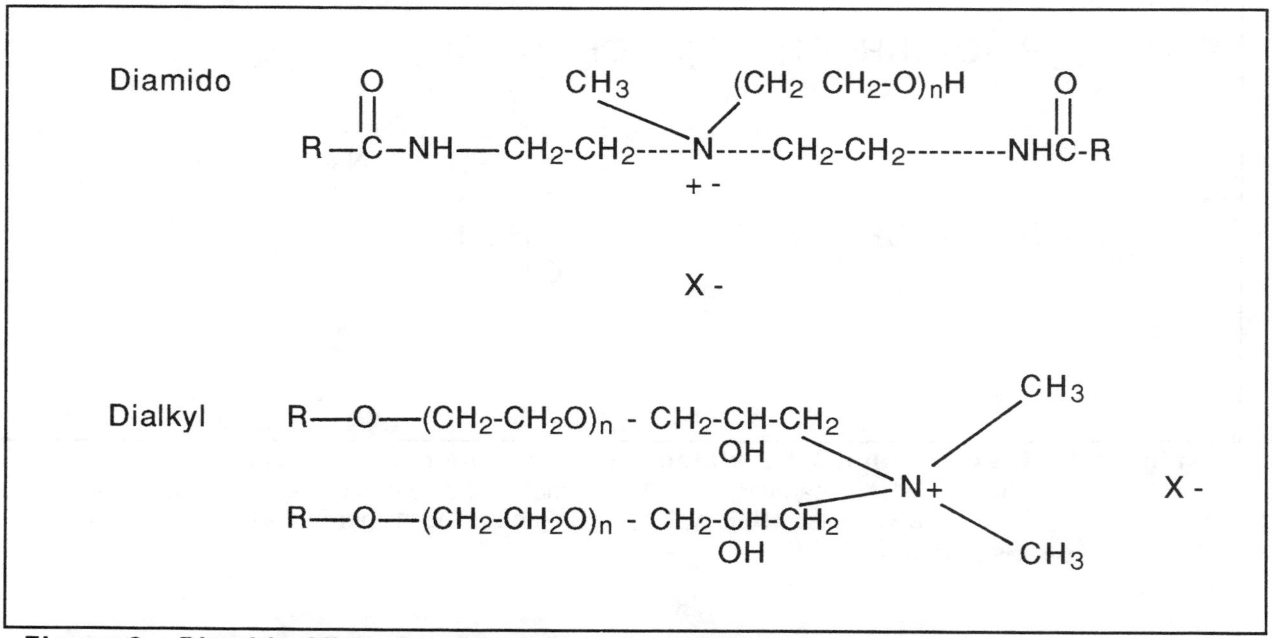

Figure 1. Dialkyl Dimethyl Quaternary Ammonium Compound

Figure 2.   Diamido Alkoxylated and Dialkyl Alkoxylated Quaternary Ammonium Compound

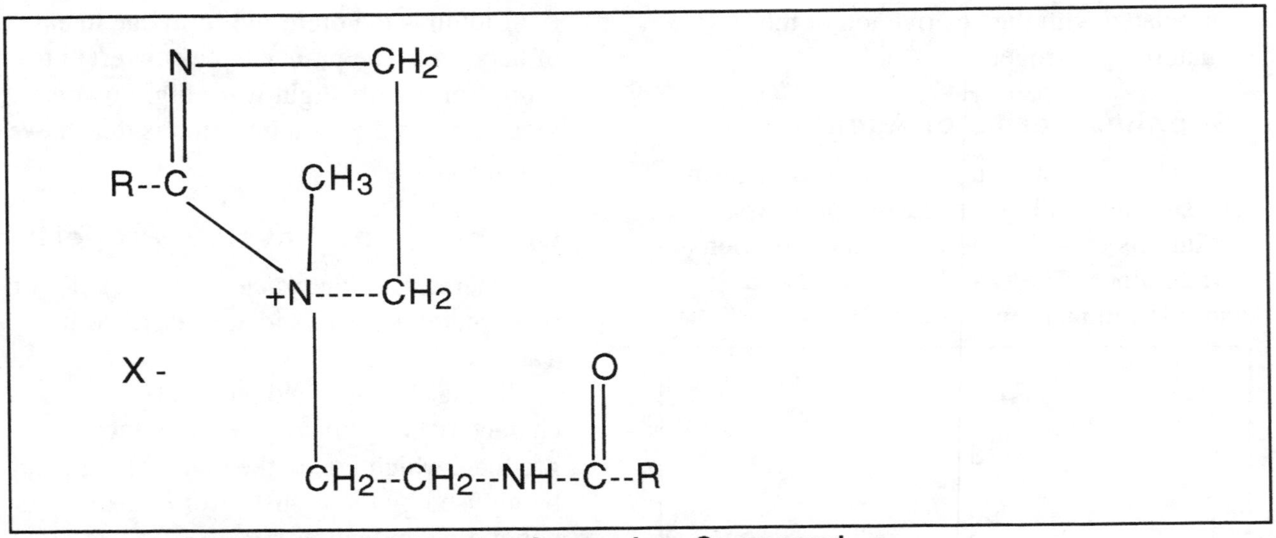

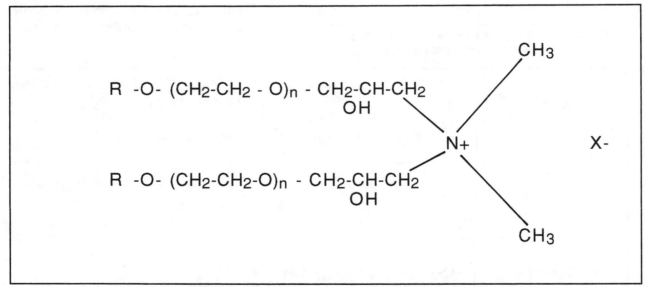

Figure 3. Imidazoline Quaternary Ammonium Compound
The fatty alkyl moiety (R) is typically C-12 to C-18 group. The other alkyl groups, which are connected by a covalent bond to the nitrogen moiety, are primarily methyl or ethyl groups. The anionic moiety is either a chloride ion or the methyl or ethyl surate ion, or both. The integer (n) has a value of 2 - 30.

Figure 4. Diakyl Phenol Alkoxylated Quaternary Ammonium Compound
The ethoxylation can vary from 9 to 30 moles of ethylene oxide/propylene oxide.
The fatty alkyl moiety (R) is typically a C9 group. The integer (n) has a value of 9 to 30 and x is a Halogen.

Table 1.   Physical and Chemical Properties of Softening and Debonding Agents

Appearance	Clear, colorless amber liquid or an amber-to-white paste
Viscosities	10 - 1000 cps
Density	0.93 - 1.10
Solids	25 - 100%
SAP value	50-155
Water solubility	Emulsifiable or dispersible
HLB	6 and up

materials in softening sanitary tissue, producing debonded pulp, or both, are:

1. Substantivity.

2. Lower dosage rate.

3. Nonfoaming nature.

4. Does not adversely affect release.

5. Not pH-sensitive.

6. Does not adversely affect brightness.

7. Does not interact with other additives in the system.

Advantages can be disadvantages, depending on the product use. Such disadvantages of using softening and debonding agents include:

1. Adverse effects on absorbency.

2. Lower production rates (debonded pulp).

Criteria for Product Selection

Criteria for product selection requires knowing various chemical types available and how variations in chemical structure influence desired properties. In review, the four types of quaternary cationic products used in the industry today are:

1. Dialkly dimethyl quaternary ammonium compounds.

2. Dimido alkoxlyated quaternary ammonium compounds.

3. Imidalazoline quaternary ammonium compounds.

4. Dialkyl phenol ethoxylated quaternary ammonium compounds.

In general, it is the fatty moiety which influences performance properties of the cationic used as a softener or debonding aid rather than the chemical type. The long-chain C-18, saturated and unsaturated fatty alkyl groups, demonstrates desired physical properties and performance. This is accomplished through the lubricating nature of the quaternary ammonium compound by reducing inter-fiber bonding, leading to greater bulk and plasticizing effect on the fiber and other additives in the system.

Shipping Containers

The standard package for cationic softeners and debonders is a 55-gallon steel or epoxy-phenolic-lined steel drum. The choice of a lined or unlined drum depends on the corrosiveness of the particular product. These products are also shipped in bulk or in plastic or steel liqui-bins containing 200-400 gallons.

Storage

Drum or Liqui-bin Storage

Storage in drums or liqui-bins should be at room temperature and away from sources of heat. Some formulations may contain solvents for viscosity control and will carry a combustible or flammable label. These products should be stored away from heat, sparks, and open flames. If outside storage is necessary, store in a dry location out of direct sunlight.

Bulk Storage Equipment

Debonding and softening agents are primarily quaternary ammonium compounds which can be corrosive to metals. Bulk storage tanks should therefore be constructed of fiberglass-reinforced resins or polyolefins. The fiberglass tank can be constructed of furan or other polyester-type resins such as the Atlac resins. Existing steel tanks can be protected by constructing a fiberglass-reinforced resin insert inside the tank. This should not be a coating since such coatings have a tendency to crack due to differences in the coefficient of expansion of the coating and steel.

Transfer lines should be of 316 stainless steel or Monel alloy. Transfer pumps and valves should be 316 stainless steel. Metallic, Viton, or Teflon seal materials are acceptable.

With flammable or combustible products, explosion-proof pumps and electrical equipment should be used.

Most debonders or softeners will titer. It is recommended that these materials be maintained at temperatures of 60-80°F (16-27°C) depending upon the titer temperature. Temperatures should be maintained below 100°F (38°C). External steam, electrical tracing of the tank, or internal heating coils of Monel alloy are suitable for maintaining temperature. Insulation of tanks with polyurethane foam is recommended. A temperature control device should be used to prevent both over and underheating.

While stirring is not required on a continuous basis, it is useful when new material is added to maintain uniform composition. The tank should have a manway, atmospheric vent, and at least one 3-inch valve for filling and one 2-inch valve for removal of the material.

Feeding Equipment and Systems

Debonders and softeners should be added to the pulp slurries as dilute emulsions in water. Figure 5 illustrates a feed system for continuous in-line dilution. Figure 6 illustrates a batch dilution system. The continuous in-line dilution system is recommended for debonding applications and most softening applications. The batch dilution system may be used where a combination of dosage and production rates would lead to very low addition rates of the neat material.

The batch dilution system consists of a 200-300 gallon tank, constructed of fiberglass-reinforced resins or polyolefins, an agitator, a chemical feed pump, and a rotometer for monitoring addition rates. The rotometer should be calibrated vs. tank draw-down while pumping against line pressure. The tank should be equipped with a level measuring device to facilitate the measurement of draw-down rates.

The addition of debonder or softener can be done manually or automatically by level controllers in the tank. When making emulsions, the cationic material should always be added to the water after the tank is

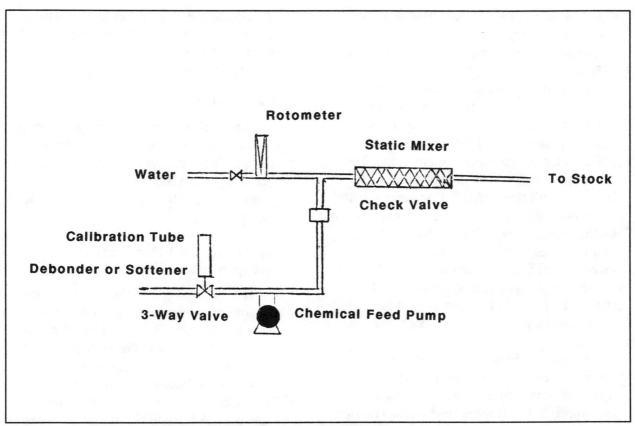

Figure 5.   Continuous In-line Dilution Feed System

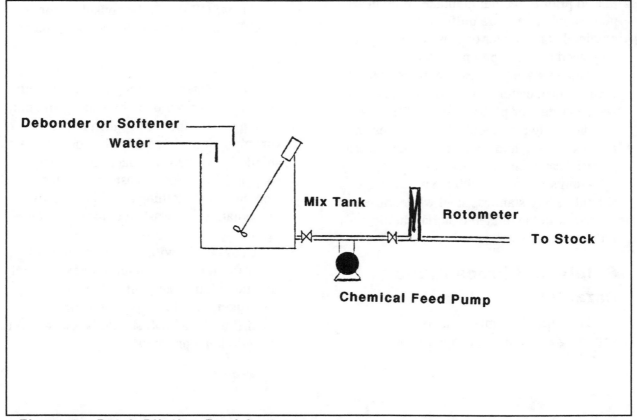

Figure 6.   Batch Dilution Feed System

filled, or as it is being filled. The best emulsions are obtained when the water temperature is in the range of 70-100°F (21-38°C).

The continuous in-line dilution system consists of a chemical feed pump which pumps the neat material into a water line. The mixture is then passed through an in-line static mixer which forms the emulsion. The dilution ratio should be at least 20:1. However, dilutions of 100:1 or greater will aid in obtaining better mixing and distribution on the fibers. In order to maintain uniform dilution rates, a rotometer should be incorporated in the water line. Control of the water line pressure is also desirable. The static mixer should be sized to give the desired mixing at the lowest water flow rate to be used.

A positive-displacement piston-type chemical feed pump should be used for flows up to 0.4 gpm (1500 cc/min). A gear pump is recommended for higher feed rates. The flow can be monitored with a calibration tube placed prior to the pump intake. With gear pumps, where a more uniform flow is obtained, a rotometer or mass flowmeter can be placed after the pump discharge.

The lines and fittings coming in contact with the neat cationic material should be 316 stainless steel or plastic. Polyolefin, PVC, or Tygon tubing are recommended. Lines and fittings coming into contact with the diluted material can be mild steel or plastic.

Pumps in contact with the neat material should be 316 stainless steel while those in contact with the diluted material should be mild steel.

Safety and Precautions
Hazards

The cationic compounds used in softening and debonding agents can be irritating to the skin and eyes. Prolonged contact can cause burns. Some people exhibit allergic reactions which should be treated accordingly.

Flammability

Some debonding and softening agents contain flammable solvents for viscosity control. These products should be handled accordingly. An MSDS for individual products should be consulted for specific hazards.

Handling and Protective Equipment

The doses of materials used in debonding and softening agents are generally irritating to the skin and eyes. Personnel handling these materials should wear eye protection consisting of either goggles or chemical face shields, as well as gloves and aprons. Disposable gloves are best to minimize exposures. Adequate ventilation is necessary to minimize vapor inhalation.

An MSDS should be consulted for specific information on individual products.

First Aid

Skin. Material coming in contact with skin should be removed by patting and not rubbing. Care should be taken to prevent spreading of the material. The contacted area should be washed with large quantities of soap and water for at least 15 minutes. Contaminated clothing should be removed immediately and washed before wearing again.

Eyes. Flush with large quantities of water for at least 15 minutes. Seek medical attention immediately after first aid.

Ingestion. Since specific formulations can differ, the MSDS should be consulted for the individual products.

Handling Spills

If the spilled material is flammable, remove all sources of ignition from the area. Soak up small spills with an absorbent material and remove to closed containers. Larger spills should be flushed with water into a suitable catch basin where they can be disposed of in a suitable manner. The contaminated area should be washed with a mild detergent and water.

Waste Disposal

The disposal of debonding and softening materials should be done in accordance with local, state, and federal EPA regulations. Since these regulations will vary depending upon location, the proper authorities and manufacturer should be consulted as to the specific regulations.

As a general rule, the cationic materials in debonding and softening agents should not be disposed of in large quantities into biological treatment ponds. High concentrations of these materials could inhibit microbiological processes in the sewage treatment process.

Application Methods and Treatment Rates
Debonder Application Methods

Debonding aids are applied as dilute emulsions in water. The emulsions can be made up in a holding tank or prepared continuously with an in-line static mixer. Debonding aids can be added at the machine chest, the fan pump, or the headbox. They should be added at a point where there is good agitation to assure adequate mixing with the fiber slurry. Addition of a highly diluted mixture to the diluted stock is preferred since this gives the best assurance of uniform distribution of the debonder on the fibers.

Application systems are illustrated in Figures 5 and 6.

Debonder Treatment Rates

The range of treatment rates required for debonding aids is usually 3-10 pounds per ton of pulp. The actual dosage rate will depend upon the type of furnish and the degree of mullen reduction required.

Softener Application Methods

Cationic softening agents are most effective when applied in the wet end of the paper machine as dilute emulsions. These dilutions can be prepared in a holding tank using warm water and high-speed mixing. Continuous in-line mixing with the use of a static mixer is also an acceptable method of preparing emulsions.

Softening agents can be added at the machine chest, the repulper, the fan pump, the headbox, and in the white water return system. Points of addition at any of these locations should be chosen with optimal agitation as a criterion for selection in order to assure good distribution of the softening agent.

This addition point should also be before the addition of any other cationic additives, such as wet-strength resins. The competition with other cationic materials for the anionic sites on the fiber can reduce the effectiveness of the softening agent.

Softener Treatment Rates

Treatment rates generally used for softening agents are 1-6 pounds per ton of pulp. Above six pounds per ton, a point is reached beyond which there may be no improvement in softness, only a decrease in strength. The treatment rate will depend to a large extent upon the type of furnish. Furnishes high in secondary fiber are generally more difficult to soften. The degree of softening that can be achieved with a chemical softening agent must be balanced with the strength requirements of the sheet.

Softener and Debonder Control Procedures

There are a number of control procedures which can be used during a softener or debonded run. These are normally determined during a laboratory evaluation or a mill trial. However, in production runs, the most rapid and accurate procedures allow operations to run at maximum proficiency.

With softening agents, tensile control methods, such as the Handleometer and the hand feel (the most subjective and reliable method as anything in practice) are typically employed.

With debonded runs, control methods center around mullen and brightness.

For both softness and debonding, the most important control factors are strength, brightness, and hand feel.

Evaluation of Effectiveness

The effectiveness of the cationic quaternary ammonium compound can be determined by appropriate test methods capable of measuring desired performance properties.

As a debonding aid, the effectiveness of a quaternary ammonium compound is determined by its mullen reduction, tensile reduction, absorbency, and brightness. In some cases, a laboratory pinwheel, hammermill, or both may be used to determine energy requirements.

For softness, effectiveness is measured by observing tensile, brightness, effect on color, and absorbency. The Handleometer and hand feel are typically employed.

Potential Problems and Solutions

Incompatibility problems can result from the presence of a cationically charged molecule in a heterogenous type system such as the papermaking process. The anionic nature of quaternary ammonium materials and other additives found in the system can react with the cationic quaternary ammonium material and render it ineffective. Other additives or byproducts of the digestion process can render a cationic ineffective. As previously mentioned, competition with other cationic additives for the anionic sites on the fiber can reduce effectiveness.

Other adverse side effects include foam, deposit formation in the system and the sheet, reduced machine speeds, and over-release. Over-release is strictly related to tissue operations where softness is desired.

Generally, the cationic materials sold today do not present handling problems. Most are liquid at 70°F (21°C) and do not present handling difficulties.

Sources of Additional Information

1. Goodbar, R. C., Riegel Textile Corp., *Multilayer Absorbant Pad for Disposable Absorbent Articles and Process for Producing Same*, U.S. Pat. 4,259,958 (April 7, 1981).

2. Haddad, Y. M., "Theoretical Approach to Inter-fiber Bonding of Cellulose," *J. Colloid Interface Sci.* 76 (2):490-501 (1980).

3. Emanuelsson, J. G., and S. L. Wahlen, Berol Kemi AB, *Quaternary Ammonium Compounds and Treatment of Cellulose Pulp and Paper Therewith*, U. S. Pat. 4,144,122 (March 13, 1979).

4. Hervey, L. R. B., and D. K. George, Riegel Textile Corp., *Method for Improving a Fluffed Fibrous Wood Pulp Batt for Use in Sanitary Products and the Products Thereof*, U. S. Pat. 3,395,708 (August 11, 1968).

5. Benz, C., Scott Paper Co., *Absorbent Fibrous Webs and Method for Making Them*, Can. Pat. 978,784 (December 2, 1975).

6. Becker, H. E., A. L. McConnel, and R. W. Schutte, Scott Paper Co., *Fibrous Sheet Material and Method and Apparatus for Forming Same*, Can. Pat. 978,465.

7. Scott Paper Co., *A Soft Absorbent Fibrous Sheet Material and a Method for Making Same*, Brit. Pat. 1,365,230.(August 29, 1974).

8. Meisel, F. W., Jr., and K. C. Larson, Scott Paper Co., *Sequential Addition of a Cationic Debonder, Resin, and Deposition Aid to a Cellulosic Fibrous Slurry*, U. S. Pat. 3,844,880 (October 29,1974).

9. Riegel Textile Corp., *Debonded Cellulose Fiber Pulp Board and Method of Producing the Same*, Brit. Pat. 1,348,409 (March 20, 1974).

10. George, D. K., and J. H. Angel, Riegel Textile Corp., *Debonded Cellulose Fiber Pulp Sheet and Method*, Can. Pat. 947, 915 (May 28, 1974).

11. Benz, C. S., Scott Paper Co., *Soft, Absorbent Fibrous Webs Containing Elastomeric Bonding Material and Formed by Creping and Embossing*, U. S. Pat. 3,817,827 (June 18, 1974).

12. Estes, P. W., Riegel Textile Corp., *Process for Forming a Fluffed Fibrous Pulp Batt*, U. S. Pat 3,809,604 (May 7, 1974).

13. Freimark, B., and R. W. Schaftlein, Scott Paper Co., *Cellulosic Sheet Material Having a Thermosetting Resin Bonder and a Surfactant Debonder and Method for Producing Same*, U. S. Pat. 3,755,220 (August 28, 1973).

14. Riegel Textile Corp., *Improvements in or Relating to Mechanically Fiberizable Cellulose Fiber Pulp Boards*, Brit. Pat. 1,282,593.

15. Champaigne, J. F., Kimberly-Clark Corp., *Manufacture of Cellulosic Fluffed Sheet*, U. S. Pat. 3,556,931 (January 19, 1971).

16. Hervey, L. R. B., and D. K. George, Riegel Textile Corp., *Method for Producing a Fiber Pulp Sheet by Impregnating with a Long-Chain Cationic Debonding Agent*, U. S. Pat. 3,554,862 (January 12, 1971).

17. Riegel Textile Corp., *Improvements in or Relating to Fluffed Fibrous Wood Pulp Batts for Use in Sanitary Products*, Brit. Pat. 1,180,801 (February 11, 1970).

18. Hervey, L. R. B., and D. K. George, Riegel Textile Corp., *Method for Producing a Fluffed Wood Pulp Batt for Sanitary Products*, Can. Pat. 809,926.

Chapter 14

Deinking

by Tom W. Woodward

Abstract

Processing aids represent a significant deinking operating cost and should be carefully selected and optimized. An understanding of deinking mechanisms and chemicals will allow optimum performance while minimizing negative interactions. It makes good economic and quality sense for mills to thoroughly investigate the wide range of chemicals and formulated products available to find the combination of processing aids that is best for their system.

Introduction

A large proportion of secondary fiber is recycled without removal of ink. However, for paper products requiring relatively high brightness and cleanliness, such as newsprint, printing and writing papers, tissue, specialty papers, and boards, the ink must be removed from the fiber and separated from the stock. Deinking operations utilize chemical, mechanical, and thermal energies to produce a fibrous suspension from printed wastepaper which is sufficiently clean and bright for use in various high quality grades of paper and board.

This chapter will briefly cover deinking systems, printing inks, and chemical mechanisms. Deinking chemical additives and their application will then be covered in detail.

The Deinking Process
System Design

A deinking system is designed according to the quality of deinked stock desired and the furnish type to be deinked. Chemical and mechanical technologies are available that will allow use of highly contaminated wastepaper in the deinking process (1-11). This involves much more than just the removal of ink. It requires handling strategies to be developed around the entire stock preparation area. Most systems involve high consistency stock handling from pulper to bleaching stages (12) (13).

A deinking system should be designed to meet the individual needs of a particular mill and possess enough flexibility to handle inherent variations in recycled fiber supply. Zabula and McCool's study on the methodology of system design is an excellent reference (14).

All systems have some basic characteristics in common. These are:

1. Removing ink from fiber (pulping)

2. Removing ink from stock
   (cleaning/screening and
   washing/flotation)

3. Bleaching

Removing Ink from Fiber

Pulping

The first stage in the deinking process is referred to as pulping or repulping. The secondary fiber is defibrated and the ink is removed from the fibers and dispersed. Defibering is a necessary operation for all secondary fiber utilization, whether or not it is to be deinked, and is covered in another chapter of this publication.

Pulping is a critical operation in deinking because in this stage ink is removed from the fiber and the particle size is most efficiently controlled. Pulping may be achieved by batch or continuous methods. The batch method is more common as it provides better control of the pulping process. Some mills prefer continuous pulping because it produces more pulp per unit. Chemicals are normally added to the pulper just prior to the addition of furnish. Pulping consistencies are usually between 4 and 6 percent. However, there is a trend toward higher consistency pulping (12-15 percent) because of the savings in chemicals, heat, and operating personnel

Pulpers have been specifically designed for secondary fiber operation (6) (15). The amount of mechanical energy generated by the pulper is important in determining the rate of defibering and the rate of ink removal and

dispersion. This mechanical energy is dependent upon the pulper configuration and pulping consistency. The chemicals added to the pulper, however, are the primary determinant of the level of ink dispersion and will be discussed in a later section of this chapter.

Removing Ink from Stock

After ink is removed from the fiber or coating, it must be removed from the stock. This is accomplished by various combinations of cleaning, screening, washing, and flotation processes.

Cleaning and Screening

Forward (conventional) centrifugal cleaners remove particulates having specific gravities greater than wood fibers. Reverse cleaners are used for removal of light contaminants. Particle size and shape have some influence on ink removal by centrifugal cleaners, with larger particles of 100-1000 micrometers being more effectively removed.

Following centrifugal cleaning, the stock is screened with either pressure screens or open vibrating screens. Ink removal by screening is poor because flat ink particles tend to align themselves with fibers and pass through the screen (16).

Washing and Flotation

After cleaning and screening, the remaining ink particles are separated from the

Table 1. Typical Commercial Washer Performance

| Washer | Consistency, % o.d. | | | | 1 Stage fiber loss, % | 1 Stage ash removal, % |
	Inlet	Discharge	Effluent w/o Ash	Effluent w/Ash		
Sidehill screen	0.6-1.4	3 - 4	0.15 - 0.25	0.22 - 0.39	12 - 18	60
Gravity decker	0.7-1.2	4 - 6	0.04 - 0.12	0.9 - 0.21	6 - 12	55
Inclined-screw extractor	3.0-3.5	8 - 12	0.25 - 0.35	0.45 - 0.73	8 - 12	45
Screw press	3.5-4.5	20 - 28	0.10 - 0.20	0.26 - 0.53	3 - 5	35

stock by washing or flotation. The advantages and disadvantages of each process are well-documented in the literature (17-28). In the flotation process, a series of flotation cells is used. A flotation cell is a tank supplied with air bubbles. A favorable chemistry in the cell causes ink particles to adhere to the air bubbles. These bubbles rise to the surface as a froth and are skimmed off as rejects.

A series of secondary cells is used to increase fiber yield. The froth from these secondary cells is concentrated for disposal and the accepts are sent back to the first primary cell. Ink removal effectiveness decreases as ink particle size falls below 40-50 micrometers. Yet, 90 percent of particles over 140 micrometers are removed (19).

Washing is a series of dilution and thickening steps repeated enough times to produce a clean pulp. Depending upon the type of washer, a wide range of consistencies can be used (18). The particular washer should be chosen relative to the type of ink to be removed, water consumption, and quality requirements of the deinked stock. Table 1 details typical commercial washer performance data for most commonly used washing devices (20).

Ink particle sizes between 2 and 20 micrometers are necessary for effective washing. Chemicals must be added to the pulper to achieve this level of dispersion. As particle size increases, the percentage of ink removed by washing will decrease. Ink particle sizes approaching 100 micrometers will not wash out even at low consistencies.

Flotation is used extensively in European deinking systems while washing has predominated in the United States. To take advantage of the benefits of both technologies, most new deinking plants feature a combination washing-flotation system. A flow diagram of a combined

washing-flotation system is featured in Figure 1.

Dispersing Units

Atmospheric dispersion units used in front of flotation and washing stages have been used successfully to reduce the size and number of ink particles. These units employ internal mechanical forces at high temperatures (non-pressurized) to reduce contaminants, such as inks and stickies, to a size which minimizes their effect on product quality and runnability and makes them easier to remove (29) (30).

Printing Inks
Components

Almost all paper inks can be subdivided into three general components: pigments, vehicles, and modifiers.

Pigments are insoluble, colored materials within the vehicle and, therefore, must be dispersed. The pigment type is determined by the desired color. Pigment supplies the proper contrast to the image area. Dyes, soluble colored materials within the vehicle, are not generally used because of their low resistance to light and a tendency to migrate into fibers. Deinkability is not normally influenced by pigment type.

The ink vehicle is the most important component in determining ease of ink removal. A vehicle is composed of a resin (binder), which binds pigment particles together and to the surface of the paper, and a solvent, which provides the ink with proper fluidity (31). Binders, when dried, form polymerized films which vary greatly in their chemical resistance.

Some common binders are listed in Table 2. Synthetic resins are used almost exclusively due to much better control of cross-linking, less polymerization necessary

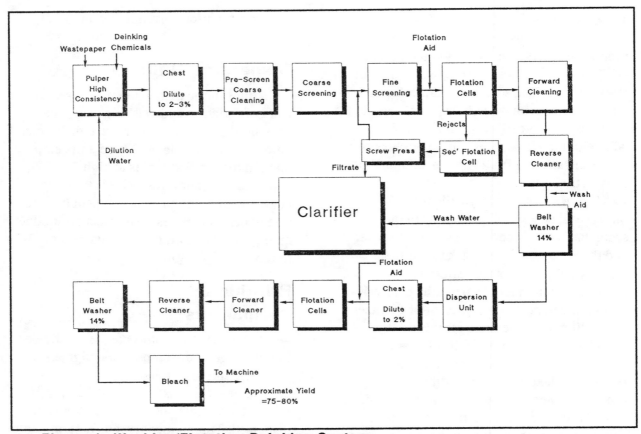

Figure 1. Washing/Flotation Deinking System

to produce a solid film, and faster-drying inks.

Modifiers are materials which give inks specific chemical or physical properties. Examples are waxes, plasticizers, drying agents, and co-solvents. They are generally added in small quantities and do not impact the deinkability of inks.

In general, offset printing carries the thinnest ink film while gravure printing carries the heaviest ink film. Multicolor prints may have 1-3 g/m^2 of ink or 34-104 pounds of ink per ton of printed waste (32).

From the perspective of the chemistry of ink removal, chemical type (vehicle) is most important. However, since chemical type is largely determined by the ink drying method, for the purposes of this discussion, inks will be classified by the method they are dried.

Types of Ink Drying Mechanisms

Ink is frequently classified according to its drying mechanism. There are four general mechanisms for drying ink: absorption, evaporation, oxidation, and radiation curing. These are listed in Table 3 (31) (33). It should be noted that many inks dry by a combination of these mechanisms (34). For example, most inks for sheet-fed offset printing contain oleoresinous or drying oil vehicles and may use a combination of absorption, evaporation, and oxidation for drying and setting of ink.

Three additional types of ink include: catalytic overprint varnishes, water-based inks for flexographic printing of newspapers, and thermoplastic dry toners used in plain paper copying and high-speed laser printers.

Table 2. Common Binders

Binder	Description
Rosin Ester	Esterification of rosin acids with glycerol or sorbitol hardened with a condensate of phenol and formaldehyde.
Petroleum Resins	Polymerization of unsaturated hydrocarbon fractions.
Alkyd Resin	Reaction of polyfunctional acids with polyfunctional alcohols condensed with drying oil fatty acids (forms oil-modified polyester resins).
Radiation-Cured	Photo or electron beam initiated free radical polymerization of epoxy acrylates, urethane acrylates, or polyester acrylates.
Water-Based	Alkaline water soluble vinyl or styrene acrylate copolymers with amine addition for alkalinity.

Absorption

In absorptive inks the vehicle (oil) penetrates into the paper surface leaving a higher viscosity pigment-vehicle blend on the surface. These are the simplest inks and are used primarily for letterpress and offset printing of newsprint and uncoated groundwood. With the possible exception of the resin binders in some offset inks, the vehicle acts like a hydrocarbon, and therefore is not saponifiable with alkali. Emulsification of the oily vehicle with dispersion of pigment particles will remove absorptive ink.

Evaporation and Heat Set

Evaporative inks dry when the solvent evaporates leaving the resin binder and pigment behind on the paper. Environmental concerns have mandated a trend toward reducing solvent concentration leading to

Table 3. Mechanisms for Dryer Ink

Drying Method	Printing Process	Vehicle Composition	Ease of Ink Removal
ABSORPTION — liquid vehicle is absorbed into paper — does not "dry" but causes an increase in viscosity of ink surface.	Letterpress and offset printing of newsprint and uncoated stocks.	•Hydrocarbon (mineral) •Hydrocarbon resins	•Not subject to saponification •Vehicle must be emulsified and/or mechanically dispersed.
EVAPORATION — includes heat set inks — volatile solvent component of vehicle is evaporated, leaving resin binder.	Letterpress, web offset, and rotogravure printing on both coated and uncoated papers.	•Hydrocarbon solvent •Rosin esters or metal resinates •Hydrocarbon resins •Alkyd resins and oleoresinous varnishes.	•Rosin esters difficult to saponify •Metallic resinates saponifiable •Hydrocarbon resins need to be emulsified and/or dispersed.
OXIDATION — solvent penetrates sheet leaving drying oil/resin combination on surface. Drying oil absorbs oxygen and film polymerizes to solid state. •Formulated to produce a quick set effect	Mainly offset sheet-fed coated and un-coated, board and paper	•High boiling hydrocarbons •Oil-modified alkyds •Oleoresinous varnishes •Phenolic-modified rosin esters	•Polymerized films not soluble in common solvents. •Partially saponified with strong alkali at elevated temperatures.
RADIATION CURING — upon exposure to radiation (UV or electron beam). Monomers in the vehicle polymerize to product a cross-linked film.	Luxury packaging, metal decorating, screen printing, protective coatings	•Epoxy Acrylates •Polyol Acrylates •Urethane Acrylates •Photo Initiators - (aryl ketones)	•Not saponifiable •Chemical dispersion difficult •Not soluble in common solvents

higher solids inks. Most web printing uses heat set inks. These are quick-drying inks formulated with high boiling petroleum oils (solvents) that vaporize rapidly when subjected to high velocity hot air or gas flow impingement.

The binder systems of these inks have become increasingly more difficult to break up and disperse for washing deinking without high alkali and surfactant-dispersant. Flotation systems, however, have little trouble removing evaporative and heat set inks.

Oxidation

Inks that dry by oxidation are usually used for sheet-fed offset printing of publication and packaging products. These inks are blends of drying oils and resins. Solvents are added to control viscosity and tack. Ink set is achieved by solvent absorption into paper. This precipitates a drying oil-resin combination onto the surface. Oxidation of the drying oil takes 1-8 hours and forms a tough, flexible polymerized film. Most of these inks can be removed from fiber and dispersed with high alkali (pH 11.5) and surfactant concentrations. However, these inks are not soluble in solvents suitable for deinking.

Radiation Curing

"Rad-cure" inks have high ink-drying rates, low or no solvent emissions, and low energy usage (35). Substrates can be printed with rad-cure inks which conventional printing cannot ordinarily handle (36). Although this latter advantage has probably helped drive the technology, it is not particularly relevant to deinking paper stock.

The two most common types of radiation-cured inks are ultraviolet (UV) and electron beam (EB) curing. UV inks dry by photo-initiated free radical polymerization (cross-

linking) of the monomer-oligomer binder. There are no solvents and the curing is very rapid. Electron beam curing does not require a photo initiator for free radical formation. Almost all of these inks are based on acrylate chemistry.

The use of radiation-cured materials as overprint coatings (as opposed to inks) appears to be a growing trend. Glossy rub-proof overlays are common in the printing of book and magazine covers and lead to difficulty in deinking since they are generally radiation-cured. There is evidence that clear UV coatings over conventional litho inks will make UV inks obsolete. At a lower cost, a clear UV overcoat will do everything five or six UV inks would do (37). These coatings are used for sheet-fed litho printing of high quality folding cartons and packaging. New application opportunities for UV and EB curing are in metallized paper, pressure-sensitive adhesives, and release coatings (38).

Catalytic Overprint Varnishes

Inks and coatings of this type utilize acid cross-linkable prepolymers of polyester-alkyds or urea. Less commonly used are melamine-formaldehyde resins combined with a solvent and a blocked acid catalyst that becomes unblocked at the drying temperature (39). These clear finishes cure after heating to 300-350°F (149-177°C) and are very difficult to deink. They show deinking resistance similar to UV coatings (40).

Catalyzed water-based coatings are beginning to compete with coatings on sheet-fed presses. Catalyzed coatings are said to be recyclable while offering higher scuff resistance (37). Currently these coatings can be used only on non-alkaline papers (40).

Water-based Inks

The relatively new water-based inks used in flexographic printing of newspapers are called publication flexo inks. These inks have only water as the carrier or volatile portion of the ink (41). The binders are alkali-soluble vinyl or styrene-acrylate copolymers and amines which are added to achieve the required alkalinity. The inks dry by penetration and evaporation. As the amine volatilizes, the acrylic resins precipitate. No organic solvents are present.

As water-based inks are alkali-soluble, they are readily dispersed and can be washed from pulp in mildly alkaline medium. They tend to re-precipitate in acid pH to form somewhat tacky masses that are difficult to wash out. The presence of alum tends to enhance this effect (42). It has been reported that these inks are not readily removed by flotation cells using conventional flotation chemistry (43). However, recent literature suggests some success in flotation removal of water-based flexo newsprint inks (44).

Laser Inks

Laser printing is a variant of the xerographic copying process. A laser printer is an internal modification of a common office copier. The image creation step and resultant final copies are identical (45). In fact, there is little difference between the inks used in laser printers and those used in standard xerographic copy machines. The industry has come to generally refer to these inks as laser inks.

The images are actually produced by dry ink particles called toner, which range in size from 8-15 microns. The pigments are similar to conventional inks, but the binders are thermoplastic resins (styrene-acrylate copolymers) that will set (cross-link) at temperatures of 300-400°F (149-204°C)

during the fusing stage of the copying process.

These resins are not readily dispersible for removal by washing deinking and are not very efficiently removed by flotation. Agglomeration and densification with subsequent removal by screening and cleaning appears to be viable (46).

The Chemistry of Deinking

The chemical mechanisms involved in washing deinking are somewhat analogous to the mechanism of laundering operations. Flotation deinking requires particle surface chemistry modifications to increase the attractive forces between ink particles and air bubbles. Table 4 lists the most important deinking chemical mechanisms.

Fiber Swelling

When dry cellulose fibers are immersed in water, there is a 56 percent increase in fiber diameter. Water moves into the fibrils to form hydrogen bonds with cellulose molecules. This breaks interfibrillar hydrogen bonds resulting in swelling of fiber. Cellulose swells more in electrolyte solutions because of the penetration of hydrated ions which require more space than water molecules (47). The breaking of interfiber bonds and swelling of fibers are important steps in deinking as they greatly facilitate loosening and removal of inks and coatings from fiber surfaces.

Table 4. Deinking Mechanisms

Fiber Swelling
Saponification
Wetting
Emulsification/Solubilization
Sequestration/Precipitation
Peptization
Antiredeposition
Dispersion

Saponification

Saponification involves hydrolysis of esters in aqueous alkali, such as a water solution of sodium hydroxide. This reaction will convert the ester into its component acid and alcohol. As previously mentioned, many of the resins used as ink binders are esters and therefore can be broken up in hot alkali solutions. This is one of the principle reactions occuring in high pH deinking of conventional offset and gravure inks. The oily vehicle in standard newsprint ink is similar to hydrocarbons and is not subject to saponification. This is also the case for the modified hydrocarbon resins often used in offset inks. Phenolic modified rosin esters can be saponified under severe conditions of pH and temperatures.

Wetting

When a liquid surface is in contact with a solid, the molecules at the interface may be more attracted to the solid than to the bulk liquid. If so, the molecules tend to spread out over the solid and the surface area of the liquid is increased. This phenomenon is called wetting. Work must be done to create this additional surface area. The work required per unit increase in area is called the interfacial tension (or surface energy). When one phase is a gas, the interfacial tension becomes the ordinary surface tension.

In deinking, with liquid (water) and solids (ink and fiber), proper wetting allows more rapid penetration of chemicals into the fiber network and ink-fiber contact area and helps ink break up and separate from fiber. Surface energies influence ink collection in flotation cells. Modification of the surface chemistries of ink particles may change their attraction to air bubbles. This will be discussed in more detail in a later section of this chapter.

Emulsification and Solubilization

Emulsification is the dispersion of one liquid phase into another to form a significantly stable suspension. Emulsification is an important chemical mechanism in deinking only when there are oils present in the ink. These inks are used in letter press and offset printing of newspapers and magazines, and they dry primarily by absorption. Adsorption of emulsifying agents (surfactants) at the oil-fiber interface releases the oil from the fiber (with the pigment particles) and forms an oil-in-water emulsion.

Solubilization, simply put, is the dissolving of substances in a medium in which they are normally insoluble. Solubilization differs from emulsification in that solubilized material is in the same phase as the solution while emulsified material is a dispersion. Solubilization may be the most important mechanism for the removal of oily inks, as it has been observed ([48]) that removal of oily soil from textile surfaces becomes significant only under conditions that favor solubilization. Surfactants will function as solubilizing agents at concentrations above their critical micelle level (the point where colloidal-sized surfactant molecule clusters are formed).

Peptization

Peptization is the conversion of an insoluble solid to a colloidal state whereby the particles are electrically stabilized. The importance of peptization as a deinking mechanism is unknown. But it probably occurs, for example, as a secondary reaction following saponification of ester-based resin binders or as ink particles are mechanically and thermally broken up.

Sequestration and Precipitation

The presence of polyvalent cations - notably calcium, magnesium, and iron - can be detrimental to the deinking process even, to a certain extent, when nonionic surfactants are used. These cations can reduce negative surface charges on both fiber and ink leading to agglomeration and redeposition. Cations also may act as linkages between negatively charged fiber and negatively charged ink particles. These ions enter the system in the water or paper stock and can be removed by sequestration (formation of a water-soluble complex) and precipitation (formation of an insoluble precipitate).

Dispersion

Dispersion is the phenomenon of adjusting the surface characteristics of particles (suspension or emulsion) to prevent reagglomeration. Dispersion is accomplished by forming either electrical, steric barriers, or both, between particles. Adsorption of negatively charged dispersing agents (surfactant or inorganic ions) onto detached ink or emulsified oil particles causes mutual repulsion and prevents agglomeration. Steric barriers arise when nonionic surfactants adsorb onto particles with their hydrophilic portions oriented out into the water thus providing a mechanical barrier to particle contact.

Antiredeposition

Antiredeposition refers to preventing the deposition of solid ink and oily particles back onto fibers. They function by sterically inhibiting the approach of ink particles to fibers and can be quite effective as wash aids in washing deinking.

Flotation Chemistry

The separation of ink and fiber in a flotation cell occurs because of their different surface energies (wettabilities). Particles forming a finite contact angle at the air-liquid interface are stable at that interface. For efficient separation the contact angle of the ink particle should be as large as possible, while those of the fiber should be kept close to zero (49). Collector chemicals selectively adsorb at the ink particle surface and cause the desired interfacial energy changes. Frothing agents are generally added to ensure stability of the air bubbles. These agents also help stabilize the particle attachment to the bubble. Well-wetted paper fiber should breach the bubble-water boundary. Consequently, paper fiber does not rise with the bubbles (50). Calcium soaps of fatty acids are the most frequently used collectors for flotation cells.

Calcium ion forms a strong bridge between the negatively charged ink particle and the negative carboxyl group of the soap molecule, making them hydrophobic, which leads to adsorption onto air bubbles present in the cell (51).

It is important to control the calcium hardness of the system. If the calcium level drops to where soluble sodium soap emulsifies the inks, very small calcium soap ink complexes will form when the stock is diluted and calcium is added prior to the flotation cells (52). If fatty acids are to be added to the pulper, calcium ion levels of 100 ppm (as $CaCO_3$) should be present. Calcium ion level in the flotation cell feed should be above 150 ppm (as $CaCO_3$). If fatty acid is not added at the pulper, it should be added, along with any extra calcium, prior to the flotation cells. Flotation is achieved by lowering the negative yield potential and increasing the hydrophobic nature of the ink particles (53).

The optimum pH for calcium-based soap collectors is 8.5. Ideally, collector soaps should be added at the same point as calcium ions forming large flocs of insoluble calcium soap. If addition levels are optimized, most of this calcium soap will be removed by the cells. Carry-over should be kept to a minimum. Collector chemical feed control based on varying plant production rates is a good way to keep feed rates optimized.

Work done at a mill in Spain revealed that addition of 2 percent clay to the pulper significantly increased the removal of 100 percent of newsprint ink by flotation (14). It was hypothesized by the authors that the fine ink particles, which were difficult to remove by flotation, were adsorbed onto the surface of the clay thereby increasing their effective particle size for more efficient removal by flotation.

Washing Chemistry

In contrast to flotation deinking, effective washing of ink particles requires as high a surface energy as possible (low contact angles) and steric and electrical barriers sufficient to prevent agglomeration and redeposition.

Ink particles should be hydrophilic for good washing and hydrophobic for good flotation. A good deink washing aid will not work as a flotation aid (regardless of foaming tendency) and may interfere with flotation efficiency. Since many existing deinking plants are adding a flotation system and new plants are using both washing and flotation, it is important that either water flows be separated, chemistries adjusted, or both, when transporting from one system to the other.

Deinking Processing Aids

Processing aids for deinking are chosen based on wastepaper, ink types, and design of the deinking system. Also of importance is the quality of stock going to the paper machine. Many mills are incorporating various percentages of deinked fiber into their final product. The amount of deinked stock that may be blended with virgin fiber is largely determined by the quality of the

deinked stock. For a specific deinking system, that quality may be controlled by careful use of deinking chemicals.

Processing aids represent a significant deinking operating cost and should be carefully selected and optimized. Many studies have been published showing the deinking effects of surfactants, dispersants, and wetting agents and their respective addition levels, feed points, and interaction with other processing aids. Given the range and complexity of the deinking mechanisms previously discussed, it is important that deinking chemicals be chosen to perform the functions required by specific ink types and system designs. However, since much information needed to properly choose such chemicals is proprietary (being the result of many years of research by chemical suppliers) it makes good economic sense for mills to thoroughly investigate the wide range of chemicals and formulated products available to find the combination of processing aids that is best for each mill system.

The following sections will discuss specific deinking processing agents in some detail. The discussions will include application differences between wood-containing and wood-free furnishes.

Table 5 lists common deinking chemicals, their function, applicable furnish types, and approximate addition levels. All safety, handling, storage, and spill control information and procedures for handling and disposal of toxic and hazardous industrial chemicals have been taken from the *Industrial Chemical Safety Manual* (International Technical Information Institute, Tokyo, Japan, 1984). All users of formulated products should refer to the MSDS supplied by the vendor for information regarding safety and handling procedures. Disposal of significant quantities of chemical additive waste will not be discussed as local, state,

Table 5.   Deinking Processing Aids

Deinking Chemical	Structure/ Formula	Function	Furnish Type	Dosage (% of Fiber)
Sodium Hydroxide	NaOH	Fiber swelling ink break-up Ink dispersion	•Wood-free grades •Groundwood	3-5 1-3
Sodium Silicates	Na_2SiO_3 (Hydrated)	Peptization Ink dispersion Alkalinity and buffering Peroxide stabilization	•Groundwood grades •Lightly inked ledger	2-4
Sodium Carbonate	Na_2CO_3	Alkalinity Buffering Water softening	•Groundwood grades •Lightly inked ledger	2-5
Sodium or Potassium Phosphates	$Na_5P_3O_{10}$ Tripolyphosphate $Na_4P_2O_7$ Tetrasodium pyrophosphate	Metal ion Sequestrant Ink dispersion Alkalinity Peptization	•All grades	0.2-1
Nonionic Surfactants	$CH_3(CH_2)_n-CH_2$ $-O(CH_2CH_2O)_xH$ Ethoxylated linear alcohols Ethoxylated Alkyl Phenols	Ink removal Ink dispersion Wetting Emulsification Solubilizing Peptization	•All grades	0.2-2
Solvents	C_{12}-C_{14} aliphatic saturated hydro- carbons	Ink softening Solvation	•Wood-free grades	0.5-2
Polymeric dispersants	$-CH_2-CH-$ $C = O$ $O- (Na^+) n$ Polyacrylate -Diisobutylene Maleic Anhydride Copolymer	Ink dispersion Antiredeposition Sequestration	•All grades	0.1-0.5
Fatty Acid (Soap)	$CH_3(CH_2)_{16}$ COONa Sodium Stearate	Ink flotation Aid	•All grades	0.5-3
Peroxide	H_2O_2	Bleach Color strip	•Groundwood grades •Some whites	1-2
Sodium Hydrosulfite	$Na_2S_2O_4$	Bleach Color strip	•Groundwood grades •Some whites	0.5-1
Chlorine	$...Cl_2$ $...OCl^-$	Bleach Color strip	•Wood-free grades	0.5-3

and federal regulations must be referred to and followed precisely.

Feed Equipment and Systems

Systems for feeding chemical additives should not be directly handled by operators. All chemicals should be stored away from work areas. Pumps should be sized and calibrated to provide proper dosage. Control switches should be placed close to work areas and designed to switch off automatically when the proper amount of chemical has been fed. Pumps should be checked routinely for calibration.

Liquids should be used whenever possible for ease of handling and better control of feed rates. Bulk systems or 200-300-gallon returnable tote bins are best as they eliminate disposal of used containers. If solids are to be used, additives are best purchased in repulpable bags which can be tossed in the pulper without being opened.

Materials for pumps, feed lines, and storage containers should be determined specifically for the chemicals to be handled. For most deinking additives, with the exception of bleaches, special materials are not required. When handling chlorine compounds or peroxides, special attention should be given to vendor recommendations.

Caustic Soda

Sodium hydroxide is one of the most important deinking chemicals for wood-free secondary fiber and is used with caution for deinking high-groundwood content grades such as newsprint and coated publication. High concentrations of alkali (pH 11.5) can either saponify, hydrolyze, or both, many ink binders and will swell fibers aiding in breaking up inks and coatings. The alkali also helps in preventing the aggregation of small ink particles into larger ones which are difficult to wash out. The inks on wood-free,

non-laser ledger, computer printout, book, and lightly printed board grades may be effectively removed and dispersed (with the use of other chemicals) at pH 10-11. Heavily printed or varnish overcoated grades may require pH 11.5 or higher.

It is unfortunate that dosages of caustic soda are expressed as a percentage of oven-dry fiber. It is the amount of hydroxide ion that is critical for deinking performance and the dosage required to achieve a given pH will vary. Sufficient caustic soda should be added to each batch to attain the desired pH. It is felt that the efficiency of many deinking plants could be increased substantially by better control of sodium hydroxide addition. Sodium hydroxide is very corrosive to animal tissue. It is not combustible, but the solid form, when in contact with moisture, may generate sufficient heat to ignite combustible materials. Caustic soda should be kept separate from acids and certain metals such as aluminum. Protective clothing and shields should always be worn when handling caustic soda.

Contaminated body areas should be washed with soap and water. Eyes should be irrigated with water.

Neutralize spills and leakage with dilute (6M or less) HCl and drain into the sewer with sufficient water.

Caustic soda may be purchased as solid flakes or pellets or as a liquid with varying concentrations. Solid sodium hydroxide is available in 50-pound bags and in drum weights up to 750 pounds. It is not recommended that solid caustic soda be used for deinking. Manual handling should be avoided due to the corrosive nature of caustic soda.

Sodium hydroxide is also a deliquescent material that causes pellets and flakes to stick together making it unusable. The ideal set-up for use of sodium hydroxide is a bulk tank system equipped to handle a 50 percent aqueous solution. Tote bins in 150-250 gallon capacities can be used when usage rates do not justify a bulk tank, but drums should be avoided because of excessive handling.

Note that 50 percent solutions of sodium hydroxide will freeze at temperatures around 40-45°F (4-7°C). Steps should be taken to prevent freezing as this could cause feed lines or pumps or both to crack and leak. Solutions of 20 percent caustic soda have much lower freezing points and should be considered where applicable. Figure 2 shows the freezing point of caustic soda solutions at various concentrations.

Soda Ash

Soda ash (sodium carbonate) is sometimes used in conjunction with sodium hydroxide. It is said to cook less harshly and produce slightly brighter pulp than caustic soda alone. It is uncommon for soda ash to be used alone because of its slow cooking time, but soda ash does provide the required alkalinity and buffers at a slightly higher pH than sodium silicate. Sodium carbonate is currently used more to make adjustments in pH than as a deinking aid.

Soda ash is normally purchased as a granular solid in bags or drums. It is caustic in nature and is an irritant to eyes and respiratory tract. Rubber gloves, safety glasses, and coveralls should be worn when handling soda ash. Flush any contaminated body areas with water.

Spills should be dissolved and diluted with water, neutralized with 6M HCl and drained into the sewer with abundant amounts of water.

Phosphates

Various phosphate compounds such as sodium (or potassium) tripolyphosphate (STPP, $Na_5P_3O_{10}$) and tetrasodium (or tetrapotassium) pyrophosphate (TSPP, $Na_4P_2O_7$) are phospate-based deinking aids. Multi-functional phosphates, possessing ink dispersion and some detergency properties, are not normally used as deinking aids but as water conditioners. These phosphates will sequester hard water cations and form uncolored complexes with iron (54).

Phosphates are occasionally added to formulated products and function similar to a builder in laundry detergents. They do provide some alkalinity but good buffering will not occur at concentrations normally used for deinking. Dosages are 0.5-1 percent based on the fiber weight added to the pulper.

Polyphosphates are available as granular solids in bags, drums, and bulk. TKPP is available as a liquid in concentrations up to 60 percent solution. Phosphates are non-combustible and moderately irritating to eyes but non-irritating to skin. Goggles should be used in high-dust areas. Spills may be flushed to the sewer with water.

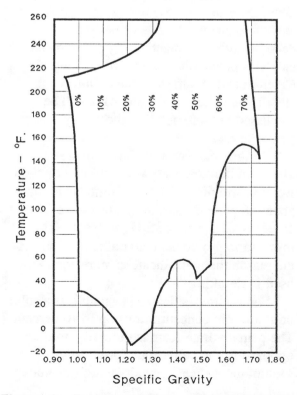

Figure 2. Freezing Points of Caustic Soda as a Function of Percent Solids

Silicates

Silicates have been used since the turn of the century in deinking wastepaper. It is said that silicates, compared to soda ash or caustic alone, provide better ink removal and brighter pulps with less fiber damage (55).

Silicates are complex solutions of polymeric silicate anions which are surface active. This surface activity (detergency) is what gives silicates many deinking functions including emulsification and suspension of dispersed ink. Deinking can occur at a lower pH than other additives.

Silicates are effective in high-groundwood furnishes and tend to cause less pulp yellowing. Silicates are used primarily in deinking newsprint or other high-groundwood furnish. They work best when small amounts of nonionic surfactants are added to aid in wetting. The sodium metasilicates are most commonly used for deinking. Additive levels are 2-6 percent.

Sodium silicate is a good stabilizer for alkaline peroxide bleaching solutions. Silicate tends to decrease the rate of peroxide decomposition by inactivating heavy metal catalysts present in the bleaching solution. Calcium and magnesium salts reinforce the stabilizing action of silicate. This will be discussed in more detail in the section on peroxide bleaching.

Commercial sodium silicate or water glass is a liquid silicate with an SiO_2 to Na_2O weight ratio of 3.22:1 and a solids level of about 38 percent. Sodium metasilicate contains a 1:1 ratio of SiO_2 to Na_2O. This lower ratio provides greater alkalinity and is the reason the metasilicate is more commonly used in deinking.

Sodium metasilicate is available as a solid or a liquid in concentrations up to 40 percent. The pentahydrate form is normally used as opposed to the anhydrous form. Solid sodium metasilicate pentahydrate is available as beads or powder in bags, drums, or bulk cans.

Silicates are non-flammable and rubber gloves, safety glasses, and protective clothing should be worn when handling. Spills should be treated as caustic and mixed and neutralized with 6M HCl and drained into the sewer with large amounts of water.

Surfactants and Dispersants

Surfactants are surface-active materials containing an organic part with an affinity for oils (hydrophobe) and another part with an affinity for the water phase (hydrophile). The hydrophobic group is usually a long chain hydrocarbon residue while the hydrophilic group is an ionic or, in the case of nonionic surfactants, a highly polar group. These surfactants function in deinking systems by lowering the surface tension of water. This enables water to wet more effectively by adsorbing onto surfaces to aid in ink removal and dispersion, and by solubilization and emulsification.

Surfactant chemistry and practical application are complex. There are thousands of available surfactants whose function and performance are influenced by many application conditions. Blends of surfactants will provide better performance than single components (48).

One useful way of characterizing surfactants is by their hydrophilic-lipophilic balance, or HLB. This is a ratio of the weight percentages of the hydrophilic and hydrophobic groups comprising a given surfactant. This ratio tends to influence dispersive and emulsifying behavior. Functional relationships of HLB are listed in Table 6 (56).

Figure 3 illustrates the effect of surfactants with varying HLB numbers on the brightness of deinked newspaper (57). Surfactants with HLB values around 14 or 15

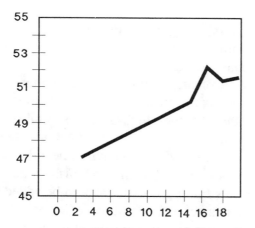

Figure 3. TAPPI Brightness vs. HLB Value

Table 6. Relationship of HLB Number to Emulsifier Use

HLB Number	Use
4-6	Water-in-oil emulsifier
7-9	Wetting agent
8-18	Oil-in-water emulsifier
13-15	Detergents
15-18	Solubilizing agents

appear to produce the highest levels of brightness.

Since there are many other surfactant properties which will influence deinking performance (e.g. chain lengths, side chains, hydrophobe type), use HLB only as a tool for screening surfactants for deinking. The nominal HLB of a surfactant can vary with application conditions, especially temperature. Two of the more common nonionic surfactants used for deinking are the ethoxylated alkyl phenols and ethoxylated linear alcohols. The hydrophilic portion of these surfactants is formed by a polyoxyethylene chain. The degree of hydrophilicity is controlled by the number of ethylene oxide units. There appears to be little difference in deinking performance between the two groups of surfactants although evidence suggests ethoxylated alcohols perform slightly better on newsprint (58). Optimum brightness of deinked ledger occurs with nine ethylene oxide units (59).

Surfactants can also be chosen for their dispersing properties which provide electrical barriers to aggregation and deposition. Recall how ink form relates to removal mechanisms. Simple newsprint ink is made up of two components, oil and carbon black pigment. Surfactants with good oil-in-water emulsifying properties will wash the oil from the fibers and release the carbon black

pigment. These pigment particles are neither emulsifiable nor bleachable and therefore must be subjected to a different mechanism which will prevent agglomeration and enhance washing or flotation, or both.

Surfactants whose primary function is to keep suspended particles dispersed are dispersants. These materials can be anionic, such as the lignosulfonates, or nonionic. Nonionic dispersants are surfactants whose characteristics (HLB, molecular weight, chain length, etc.) are such that they are not particularly effective emulsifiers or wetting agents but do provide effective electrical or steric barriers to particle agglomeration or deposition.

Polymeric Dispersants

Some hydrophilic polymers are good dispersing agents but are not classified as surfactants. Hydrophilic polymers are water soluble, multi-functional polyelectrolytes. Besides particle dispersion, these polymeric dispersants sequester cations, inhibit scale, and exhibit some detergent properties (60). Two common examples are polyacrylates and disobutylene maleic anhydride copolymer.

Hydrophilic polymers are normally blended into formulated deinking products as they tend to enhance the performance of emulsifiers and wetting agents. They are relatively expensive and optimum dosage is dependent upon the chemical environment of the deinking system.

Antiredeposition agents are materials that will prevent the redeposition of ink particles onto fiber surfaces. This is one of the basic operations in the detergency of textile materials (61) that also has relevancy in deinking. Antiredeposition agents are normally blended with surfactants and dispersants in a formulated deinking product. Antiredeposition materials are anionic or nonionic polymeric dispersants that are most commonly derivatives of cellulose such as sodium carboxymethyl cellulose (CMC) and hydroxypropyl methyl cellulose (HPMC). Hydrophilic CMC, which is a polyelectrolyte, adsorbs onto cellulose fibers by hydrogen bonding and repels inks by electrostatic repulsion (62).

Dosage levels of surfactants and dispersants can vary greatly depending upon the chemical type and deinking conditions. Levels usually fall between 0.25 and 1.0 percent based on fiber weight. Most are liquids but some can be purchased in granular or solid form. Because of the wide variety of surfactants and dispersants and blends, the user should rely on the MSDS provided by the supplier for information regarding safety, handling, storage, spill control, and disposal.

Solvents

Solvents are available that will dissolve most of the inks and varnishes in use today. Unfortunately, high costs prohibit their use in most deinking programs. Also, in order to function properly in the pulper, a solvent should be immiscible in water at rather low concentrations (<1000 ppm). Many good ink solvents are miscible with water at this concentration and, therefore, solvent-water emulsions cannot be formed. This decreases the solvating power of the solvent. Environmental concerns also limit the use of many effective solvents such as the chlorinated hydrocarbons. Because of these

factors, aliphatic hydrocarbons are the most common solvents used in deinking systems today.

Aromatic hydrocarbons are better solvents for most inks, but it is felt by this author that their higher water solubility negates their greater solvent action. Their use is also limited due to environmental concerns.

A surfactant with good oil-in-water emulsifying properties should be added when using a solvent to ensure sufficient emulsification of the solvent in the pulper. Water-miscible solvents, such as some glycol ethers, are not very effective deinking agents.

Although many binders used in inks and varnishes are not soluble in the solvents that are practical for deinking, many of these binders can be softened by a solvent. This allows easier breakup and dispersion by the mechanical action of the pulper and surfactants and dispersing agents that are added with the solvent. Dosage levels are 0.25-2 percent based on fiber weight added to the pulper prior to fiber addition.

There are a variety of aliphatic hydrocarbons solvents. Here again, environmental, insurance, and transportation constraints limit the types that may be used to those with chain length and molecular weight giving a flash point of greater than 140°F (60°C). Relatively new volatile organic carbon (VOC) emission restrictions are also limiting use of solvents. It is expected that conventional solvents as functional deinking agents will be phased out of use by the mid-1990s.

Aliphatic hydrocarbon solvents are available in drums, tote bins, and tankers. They are volatile and classified as combustible material by the U. S. Department of Transportation. They should be stored in areas with good ventilation and kept away from all sources of heat, sparks, and open flame.

Empty containers will contain solvent residues and care should be taken when handling them. Although it is not particularly toxic, prolonged or repeated inhalation or contact with skin should be avoided. Affected areas should be washed thoroughly with soap and water.

If spilled, solvents must be kept away from sources of ignition and flushed into a suitable retaining area with water. An appropriate absorbent may be used to absorb small amounts.

Fatty Acids

Fatty acids are the primary flotation collector chemicals used in Europe. In the United States, there appears to be a trend toward synthetic collectors because of the new flotation deinking systems.

Fatty acids are composed of straight or branched chain alkyl groups with a carboxylic acid functional group. The chain length typical for flotation deinking is C_{14} to C_{18}. One of the most common is oleic acid. The ionic form of the fatty acid is required for precipitation with calcium ions and effective flotation. Ionization can be achieved by adding the free fatty acid to the pulper at an elevated pH of 9.5-10.5. This hydrolyzes the free acid to its carboxylate anion. If addition just prior to the flotation cells is desired, the sodium or potassium salt (soaps) of the fatty acid should be used as this will dissolve in water to form the required anion. These anions, precipitated by calcium cation, form the reactive flotation species. Calcium cations and the soap should be added at the same point just prior to the flotation cells. Addition of calcium should be carefully controlled to convert all fatty acid anion to the calcium salt. The pH should be maintained at 8.5 or above to achieve foam stability. Non-precipitated soaps will function as frothers (17). Addition levels normally range from 0.75 to 2 percent.

There is evidence that economical addition rates are greatly temperature dependent with highest addition level of 3 percent occurring at 68°F (20°C) (19).

Fatty acid collectors are usually purchased as 50 percent solutions of the sodium salt (soap). In this form they are added (with calcium chloride if necessary) just prior to the flotation cells where good mixing is assured. Some of these soaps require storage and feed line temperature of 122°F (50°C) to avoid precipitation. Some mills will purchase solid fatty acid or soap and use a makedown unit supplied by the vendor. Handling and storage is determined by the physical properties of the collector. For example, oleic acid is a liquid at 68°F (20°C) but freezes at 55°F (13°C) and sodium stearate is freely soluble in hot water but much less so in cold. The MSDS supplied by the vendor should be consulted for safety precautions.

Synthetic Collectors

Non-fatty acid flotation collector chemicals have recently been made available. These are easier to control and perform at much lower dosages (63). These synthetic collectors are ionic or non-ionic surfactants that do not require calcium ion for ink collection. The usual dosage is 0.1-0.3 percent and may be added to the pulper or just prior to the flotation cells depending on the design of the deinking system.

Recent evidence (64) suggests that $CaCl_2$ reduces the final brightness of deinked newspaper by four brightness points and increases the consumption of peroxide. This could lead to an increase in the use of synthetic collectors. Studies have also shown that water-based flexographic newsprint inks are more effectively removed by flotation cells when synthetic collectors are used, especially at a lower pH of 7 (44). All synthetic collectors have proprietary formulas. Information regarding safety, storage, handling, and application technology should be obtained from the vendor.

Bleaching Agents

When higher brightness of deinked stock is necessary, bleaching is required. The chemicals used are determined by the type of furnish to be bleached and the brightness requirements of the end product. Bleaching can be done in pulpers, storage chests, or towers. Pulpers, however, are generally operated at consistencies that are not optimum for bleaching. Retention times are relatively short and bleaching chemicals will be uselessly consumed by materials that will be removed from the pulps in subsequent cleaning and washing stages. Brightness, however, can be increased by adding bleaching chemicals to the pulper. Ideally, bleaching should take place in bleaching towers under conditions optimal for the particular bleaching chemical.

Lignin preserving chemicals are necessary for bleaching high-yield fibers such as deinked newspaper. Chromophoric (light absorbing) groups on residual lignin can be eliminated with reducing or oxidizing agents. The two most common lignin-preserving bleaching agents are hydrogen peroxide, an oxidizing agent, and sodium hydrosulfite (dithionite) which is a reducing agent.

Chlorine compounds are oxidative lignin-removing bleaching agents. Chlorine and hypochlorite are used to bleach chemical pulps; but chlorine is not suitable for bleaching high yield fibers such as deinked newsprint. There is evidence in the literature (22) (65) that changing the oxidation state of ink components may make them easier to remove by changing their solubility rate or ease of saponification. Mild oxidizing or

reducing agents could also accomplish this, but these reactions are not part of the bleaching process.

Hydrogen Peroxide

Hydrogen peroxide (H_2O_2) is most commonly used in bleaching deinked newsprint. It is an unstable weak acid and requires careful application for optimal performance. Hydrogen peroxide disassociates in water to produce peroxide anions (HO_2), which are the active bleaching species. This occurs readily only under alkaline conditions. Since the peroxide anion is readily decomposed in very alkaline solutions to form oxygen (O_2), it is important to maintain a pH of 9-10.5.

Decomposition of hydrogen peroxide under bleaching conditions is catalyzed by small amounts of iron, copper, and manganese. This can lead to very significant losses (up to one-third) in oxidative capacity. Kutney (66) studied 75 compounds and found that sodium silicate was the most efficient for stabilizing hydrogen peroxide. While the mechanisms of hydrogen peroxide decomposition are not completely clear, the primary cause appears to be the catalytic effect of trace amounts of transition metals such as iron, copper, and manganese (67).

Hydrogen peroxide decomposition is completely inhibited by the addition of small amounts of the pentasodium salt of diethylenetriamin pentacetic acid (Na_5DTPA) (68). Concentrations as low as 0.034 g/L completely stabilized the hydrogen peroxide solution (see Table 7) (68). Note that complete removal of iron and copper from bleaching systems is not achieved by chelation (69).

Sodium silicate is an irreplaceable additive in the peroxide bleaching of mechanical pulps. Evidence (70) indicates that silicate does not stabilize peroxide by itself but inactivates hydrous iron oxides and manganese ions that catalyze peroxide decomposition during brightening. Silicates also function as a buffer to keep the pH lower at a given alkali content which also minimizes decomposition (71).

Magnesium, as magnesium sulfate, has also been used in conjunction with silicates as a peroxide stabilizer. When the two materials are added together, a synergistic effect causes a brightness gain (72). Magnesium may stabilize peroxide by interrupting the decomposition reaction rather than inactivating transition metal catalysts. In fact, magnesium is an ineffective stabilizer in the presence of iron and copper and becomes a catalyst in the presence of manganese (73).

The details of peroxide decomposition and stabilization are complex and outside the scope of this text. These recent studies, however, suggest that the traditional method of peroxide bleaching of deinked newspaper additions of peroxide, silicates, caustic, magnesium sulfate, and chelants could be improved by detailed water analyses and subsequent control of transition metals. Stabilizers could then be chosen based upon their ability to inhibit the catalytic activity of the specific metal ions present in the system.

Typical conditions for hydrogen peroxide bleaching of deinked newsprint are listed in Table 8.

Brightness increases 8-10 points under these conditions. The effects of bleaching consistency on final brightness is given in Table 9 (74). Bleaching under conditions similar to those listed in Table 8 at 20 percent consistency increased the brightness by less than 2 points over the brightness at 8 percent consistency. Brightness gains will drop considerably to only 3 or 4 points if consistencies are lowered to 3-4 percent. More brightness can be achieved by adding peroxide in a bleaching stage following ink removal (64).

Table 7. The Effect of Na₅DTPA Concentration on the Decomposition of 0.098M H₂O₂*

Na_5DTPA Concentration g/L	H_2O_2 Decomposition, %
1.700	7.6
0.340	1.5
0.170	0.7
0.034	0.0
0.000	29.0

*(Initial pH 10.8, 50 °C, 120 min)

Table 8. Typical Conditions for Hydrogen Peroxide Bleaching Deinking Newsprint*

Material	Wgt. % on Paper
H_2O_2	1
NaOH	1.5
Silicate	3
$MgSO_4$	0.05
Na_5DTPA	0.3

*(50 °C, 1.5 hr treatment)

Peroxides may also aid in the breakup of cured alkyd resin binders that are common in offset inks. The cross-linking oxygen bridges of these cured resins are oxidized by the peroxide into unstable organic peroxide bondings which decompose and break (31) (32).

Hydrogen peroxide is available in drums, tank trucks, and tank cars. A 50 percent concentration is typical, with 70 percent and 35 percent concentrations also available. Details regarding unloading and storage facilities, materials of construction for pipe lines, and valves and pumps must be obtained from the vendor because many common engineering materials must not be used with hydrogen peroxide. In general, aluminum alloy or 300 series stainless steel is acceptable.

Hydrogen peroxide is not considered an explosive but can react with other substances (metals) to form explosive mixtures. Solution

Table 9. Effect of Consistency on Bleaching Deinked Newsprint

Bleach Conditions

H_2O_2 % Consistency as noted
49 °C
3 hours

Experiment number	1	2	3	4	5	6	7	8	9	10	11	12	13	14
Consistency, %	8	8	8	8	8	12	12	12	12	20	20	20	20	20
Bleach Chemicals*														
H_2O_2, (100%)	1.0	1.0	1.5	1.5	2.0	1.0	1.5	1.5	2.0	1.0	1.0	1.5	1.5	2.0
NaOH	1.25	1.5	1.5	1.75	2.0	1.25	1.25	1.5	2.0	1.0	1.25	1.25	1.5	2.0
Na silicate	5.0	5.0	5.0	5.0	5.0	5.0	5.0	5.0	5.0	5.0	5.0	5.0	5.0	5.0
$MgSO_4$	0.05	0.05	0.05	0.05	0.05	0.05	0.05	0.05	0.05	0.05	0.05	0.05	0.05	0.05
Residual Chemicals**														
H_2O_2	20	14	23	19	24	12	20	18	17	6	7	11	9	8
Total alkali	25	27	25	29	28	20	14	18	22	8	16	9	11	7
Brightness, Elrepho**														
Unbleached	53.4	53.4	53.4	53.4	53.4	53.4	53.4	53.4	53.4	53.4	53.4	53.4	53.4	53.4
H_2O_2	64.2	64.0	66.6	66.5	67.4	65.2	67.2	67.5	68.8	66.1	65.9	68.3	68.5	70.1

* Percent chemical added on O.D. pulp.
** Percent of chemical charged; total alkali expressed as NaOH.
***Single stage hydrosulfite brightness: 60.9.

concentrations above 65 percent may ignite combustible materials. Containers should be kept separate from combustible, organic, or easily oxidizable materials and should be vented and stored in a cool, well-ventilated area. Rubber gloves, goggles, a respirator, and protective clothing should be worn because concentrated solutions will irritate skin and eyes. Spilled hydrogen peroxide should be flushed and diluted with water.

Sodium Hydrosulfite (Dithionite)

Sodium hydrosulfite ($Na_2S_2O_4$) is the most common type of reductive bleaching agent. It works effectively on deinked newsprint and white grades.

During bleaching, hydrosulfite reacts to form sulfites (SO_3^{-2}). Since the reaction consumes hydroxide ions (OH), buffering may be necessary to maintain proper pH of 5.5-6.5. Hydrosulfite also reacts with water to produce thiosulfate ions ($S_2O_3^{-2}$). Although this reaction has little influence on the bleaching ability of hydrosulfite, thiosulfate can contribute to corrosion.

Dithionite is consumed (oxidized) by the oxygen present in air; this reaction is accelerated by increasing pH. The optimum pH for hydrosulfite bleaching would be 8-9 if oxygen could be eliminated. However, pH 5.5-6.5 is the normal range for deinked pulp to keep air oxidation to a minimum.

As with peroxide, heavy metals will catalyze these decomposition reactions. Complexing agents such as DTPA and EDTA should be added as stabilizers, especially if iron content is high (>0.3 ppm). STPP is also effective at a 0.3 percent addition level.

Maximum hydrosulfite effectiveness is reached at an addition level of 1 percent and a

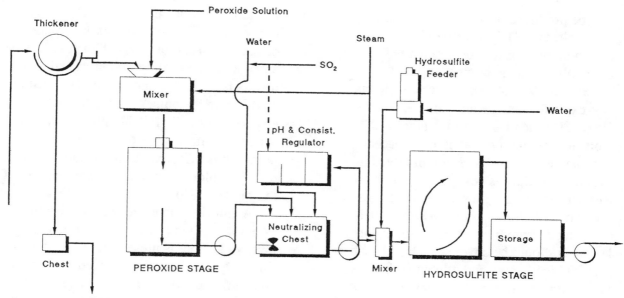

Figure 4. Two-stage Peroxide/Hydrosulfite Bleaching System

bleaching consistency of 4-5 percent. Temperatures of 120-180°F (49-82°C) will produce good results. Reaction times are very temperature-dependent ranging from 30 minutes at 180°F (82°C) to 2 hours at 120°F (49°C) (75) with 150°F (65°C) recommended. Inadequate mixing of hydrosulfite into the pulp slurry will greatly reduce bleaching efficiency.

By eliminating air during the mixing, retention, and rapid mixing processes, brightness will increase 8-10 points. Brightness increases of 4-6 points are realistic when hydrosulfite is added to the pulper. Steam added to the stock prior to bleaching will liberate air and help prevent air oxidation. The pulper can be shut down for 15-20 minutes after hydrosulfite addition to reduce consumption by air mixed into the stock by the pulper rotor (76). General conditions for hydrosulfite bleaching are listed in Table 10.

Hydrosulfite can be used with chlorine compounds to strip color (whiten) when deinking white fiber. Dyes and pigments which cannot be oxidized can often be decolorized with hydrosulfite. It can also be used to brighten pulp following a single stage of hypochlorite. Less expensive sodium bisulfite should be added prior to the hydrosulfite to reduce all residual hypochlorite.

Brightness gains of 10-14 points are possible with a two-stage bleaching sequence, wherein peroxide is added and followed by sodium hydrosulfite (77). A two-stage peroxide-hydrosulfite bleaching system is illustrated in Figure 4 (75). Bleaching conditions of each stage are similar to those previously discussed. Sulfur dioxide or bisulfite is necessary to reduce residual peroxide prior to the hydrosulfite stage.

Sodium hydrosulfite is available in powdered form (70-75 percent active) or as a liquid of 13 percent concentration. A 33 percent suspension is available (78). It can

Table 10. Conditions for Hydrosulfite Bleaching

Material	Wgt. % on Paper
$Na_2S_2O_4$	1%
STPP	0.3%
pH 5.5-6.5	
130 °F-160 °F	
30-90 min.	

also be produced on-site from sodium borohydride (79). The vendor should be consulted regarding proper construction materials for lines, pumps, and tanks.

Sodium hydrosulfite is flammable but non-explosive and decomposes at 131°F (55°C). It generates heat in contact with moisture and air and may ignite adjacent flammable substances. When involved in a fire, toxic sulfurous acid gas is generated. Rubber gloves, goggles, a respirator, and protective clothing should be worn when handling sodium hydrosulfite. Spills and leakages should be covered with soda ash or sodium bicarbonate and diluted with water into a large container. An equal volume of calcium hypochlorite should be added. After one hour, dilute and neutralize with 6M HCl or NaOH and drain into the sewer with a large quantity of water.

Chlorine Compounds

Chlorine (Cl_2) and sodium hypochlorite (NaOCl) are the two most common bleaching agents for white (wood-free) paper deinking. Both are delignifying agents and, as such, are not practical for use in bleaching of wastepaper having a groundwood content of greater than 5 percent.

It is possible to produce a lignin-free deinked pulp having a brightness greater than 80 from wastepaper containing up to 50 percent groundwood (76). Wastepaper is cooked in the pulper with 4 percent caustic at 190°F (88°C). After deinking, the pulp is subjected to a standard three-stage bleaching operation (CEH) or a two-stage process (chlorination followed by overneutralization with caustic) to produce sodium hypochlorite.

Using previously bleached wastepaper, a single-stage hypochlorite bleaching of deinked pulp will produce a brightness of 80 or more when normal levels of 2-5 percent

available chlorine are added. Calcium hypochlorite may be used but it is more difficult than sodium hypochlorite to handle and may cause scale problems. Feeding to towers is preferred and is usually continuous. Addition to the pulper is possible but it is not efficient and will probably produce a brightness increase of only three or four points. It is effective, however, in stripping residual color and will aid in breaking up of ink binders (80).

A chlorination stage is necessary for high brightness when greater than 10 percent of high-yield fiber is present in the wastepaper. The technology is essentially equivalent to conventional pulp bleaching. Given the current environmental concerns over dioxin and chlorinated organic materials, it is doubtful chlorination will be viable as a bleaching process for new deinking plants. Hypochlorite is now under investigation with respect to the same environmental concerns. Perhaps the peroxide bleaching technologies developed for high brightness bleaching of CTMPs will have some applicability for deinked pulps.

Sodium hypochlorite is available as a liquid in concentrations up to 20 percent with 15 percent being the most common. It can be purchased in drums, tote bins, or bulk loads. The vendor should be consulted for storage and handling procedures as there are many restrictions on materials of construction for feed lines, valves, tanks, pumps, etc. The shelf life of sodium hypochlorite is only a few months (shorter at higher temperatures) as it decomposes to form sodium chloride.

Sodium hypochlorite should be stored in a cool, dark place away from combustible materials in a well-ventilated area. Rubber gloves, face shields, and protective clothing should be worn when handling sodium hypochlorite.

Contaminated body areas should be washed with soap and water. Eyes should be irrigated with water. Spills should be covered with weak reducing agents such as bisulfite with 3M H_2SO_4 for faster reaction. The mixture should then be transferred to a large container of water, neutralized, and drained into the sewer with a large quantity of water.

Evaluation of Deinking Performance

Handsheet or pulp pad brightness and an estimate of dirt or specks are the most common methods in evaluating deinking performance. Other sheet properties that may be important are color and ash content.

For proper evaluation of deinking performance, the objectives of the deinking program must be well-specified. It is important that the test methods used in the evaluation adequately reflect the performance of the deinking operation. Brightness and dirt measurements are relatively easy to perform, but both require the making of handsheets or pulp pads.

Handsheet making is, in effect, a high-dilution washing stage. The loss of fines and ink during sheet preparation may obliterate the quantitative analysis of the particular step being evaluated. R. P. Cruea, in a paper on laboratory evaluation of deinking performance (81), recommends either making

pulp pads after diluting to 1 percent consistency or vacuum draining the pulp on a 60-mesh screen at testing consistency without further dilution. These procedures provide close approximations of the condition of the pulp at a particular stage in the deinking operation.

TAPPI Test Method T213 os-77 "Dirt in Pulp" is an adequate method for quantifying ink specks remaining in the pulp. The size (area) of a dirt speck on a sample is determined by comparing it with standard reference specks on the TAPPI dirt estimation chart. Dirt is reported as square millimeters of equivalent black area per square meter of surface examined (ppm).

Another evaluation method is to count the number of specks (ink particles) on both sides of the sample and divide by the weight of the sample. Results are reported as specks per gram of fiber. A measure of both TAPPI dirt and specks per gram may be used as a measure of the degree of ink dispersion.

Ink dispersion in a laboratory pulper may be measured by adding small pieces (1 inch x 1/2 inch) of polyethlylene to the pulp shortly before the end of the cook (82). A clean and ink-free plastic sheet will indicate good ink dispersion while an ink-coated plastic sheet will indicate poor ink dispersion.

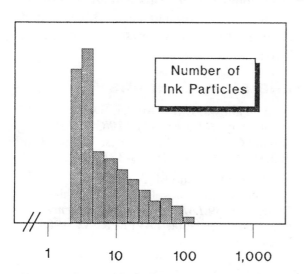
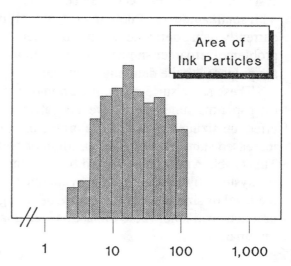

Figure 5. Size Distributions of Ink Particles for Samples of 100% Waste Newsprint Furnish with Strong Dispersion Chemistry

Comparison of curves generated by measuring reflectance at all wavelengths of visible light are more useful than brightness measurements. Curve comparisons can help determine factors contributing to a lack of whiteness by identifying specific colors being absorbed.

Image analysis is a new technique for evaluating the performance of a deinking system. Image analysis techniques allow measurement of not only the amount of ink but also the nature of the ink particle size distribution (83). Figure 5 (14) illustrates the information on ink particles obtained from an image analysis.

Adverse Side Effects

There are a number of potential negative effects of deinking chemicals both within the deinking plant and as a result of carryover into the paper machine system. Among these are foaming, scale, sizing interferences, and corrosion.

Foaming may be a problem when surfactants are used as deinking aids. In general, the most effective surfactants for deinking also have the greatest tendency to foam (81). Proper blending of surfactants will minimize the foam problem while maintaining deinking efficiency. In most systems, this type of foam may be controlled with small amounts of defoamer. Many mills currently using defoamer exhibit no foam problem. Defoamer should not be added to the pulper with the deinking chemicals.

These same surfactants, if carried over to the paper machine, could have a negative effect on sizing and may cause problems with increased starch, coating penetration, or both. This problem may be alleviated by isolating the system, by adding size, or by increasing the level of size addition. There are deinking products available where impact on sizing is minimal.

Calcium ions can enter the system as water hardness, as a flotation aid, or in the secondary fiber. It may cause deposition on side hill screens and auxiliary equipment. Treatment for calcium carbonate deposition involves the addition of a suitable precipitation inhibitor (84).

Residual chlorine carryover to the paper machines should be avoided because it will contribute to corrosion rates. Reduction of residual chlorine to chlorides with sodium bisulfite is effective.

Thiosulfate levels should be monitored especially when bleaching with hydrosulfite. Thiosulfate can cause severe corrosion even on stainless steel. Thiosulfate corrosion rates can be reduced with peroxide treatment (85).

Another problem area in deinking systems which is not necessarily related to chemical effects is what to do with a bad batch. Sometimes a system designed to handle specific types of secondary fiber and inks is disrupted by the inclusion of furnish that cannot be deinked (lasography) or repulped (wet-strength paper). It may be possible to adjust pulper operating parameters such as chemical levels, pH, time, and temperature to account for these troublesome materials in the pulper. It is also advisable to have a bad batch chest in which to dump the unacceptable stock. This will allow further chemical treatment and a careful bleeding of the bad furnish into the system at acceptable levels.

Literature Citations

1. Fallows, J. D., "Modern Deink Technology: A Case Study," *1983 TAPPI Pulping Conference Proceedings*, Atlanta: TAPPI PRESS, 1983, Book 1: 379-381.

2. Scott, J., and M. A. McCool, "Desencrage Cascades: Deinking of White Grades," *1989 TAPPI Pulping Conference Proceedings*, Atlanta: TAPPI PRESS, 1989, Book 1: 373-380.

3. Pfalzer, L., "Flotation Deinking and Waste Paper Recycling Systems," *1989 TAPPI Pulping Conference Proceedings*, Atlanta: TAPPI PRESS, 1989, Book 1: 43-47.

4 . Flynn, P. J., and M. E. Rodda, "Fine Tuning of Flotation Deinking," *1982 TAPPI Pulping Conference Proceedings*, Atlanta: TAPPI PRESS, 1982, Book 1: 453-455.

5. Woodward, T. W., "Chemicals Used in the Deinking of Secondary Fibers," *1986 TAPPI Chemical Processing Aids Seminar Proceedings*, Atlanta: TAPPI PRESS, 1986, pp. 55-62.

6. Koffinke, R. A., "Pulping Developments in the Secondary Fiber Field," From an internal technical article, Thermo Electron, Auburn, Mass.

7. Koffinke, R. A., "Energy Considerations in Deinking Systems," *Tappi J.*, 64(10):69-72 (1981).

8. Gilkey, M., "Use of Cold Dispersion, Flotation and Washing in a Japanese Deinking System," *1981 TAPPI Pulping Conference Proceedings*, Atlanta: TAPPI PRESS, 1981, Book 1: 133-142.

9. Koffinke, R. A., "Improved Deinking Systems Yield Cleaner Stock, Reduced Effluent," *Pulp & Paper*, 55(7):159-162 (1981).

10. Halonen, L., and R. Ljokkoi, "Improved Screening Concepts," *1989 TAPPI Pulping Conference Proceedings*, Atlanta: TAPPI PRESS, 1989, Book 1: 61-66.

11. Gilkey, M. W., and C. E. McCarthy, "A New Device for High Efficiency Washing of Deink Furnishes," *1988 TAPPI Pulping Conference Proceedings*, Atlanta: TAPPI PRESS, 1988, Book 3: 649-654.

12. Fisher, H. S., "Report on a High-Consistency Deinking Process for Mixed Secondary Fibers," *Paper Trade,* 164 (17):54-57 (1980).

13. Keszthelyi, S. E., and J. Babinko, "High Consistency Deinking: The Flexible Process," *1981 TAPPI Pulping Conference Proceedings*, Atlanta: TAPPI PRESS, 1981, Book 1: 283-297.

14. Zabula, J. M., and M. A. McCool, "Deinking at Papelera Peninsula and the Philosophy of Deinking System Design," *Tappi J.*, 71(8):62-68 (1988).

15. Bahr, T.,"New Developments in Repulping Waste Paper," *1980 TAPPI Pulping Conference Proceedings*, Atlanta: TAPPI PRESS, 1980, Book 1: 203-208.

16. Ekstrom, R., "Optimization of the Boise Cascade Vancouver Mill Secondary Fiber Deinking Operation," From TAPPI Pacific Section Meeting Notes, March 1982.

17. Ortner, H. E., "Recycling of Papermaking Fibers: Flotation Deinking," *TAPPI Monograph-1981*, Atlanta: TAPPI PRESS, 1981.

18. Horacek, R. G., "Recycling of Papermaking Fibers: Deinking by Washing," *TAPPI Monograph-1983*, Atlanta: TAPPI PRESS, 1983.

19. Marchildon, L., M. LaPointe, and B. Chabot, "The Influence of Particulate Size in Flotation Deinking of Newsprint," *1988 CPPA Annual Meeting Proceedings*, 1988, B: 61-66.

20. Horacek, R. G., "Using Less Water in Deinking," *Tappi J.*, 62(7):39-42 (1979).

21. McKelvey, H.E., "The Design and Use of Side Hill Washers," From an internal publication of Kalamazoo Tank and Silo Co., Kalamazoo, Mich.

22. Anonymous, "The Continuing Development of Deinking," *Paper*, Aug. 4, 1980.

23. McHela, J., and A. J. Felton, "Shortcutting the Deinking and Bleaching Process for High-Grade Secondary Fiber," *1973 National TAPPI Conference Proceedings*, Atlanta: TAPPI PRESS, 1973.

24. Horacek, R. G., and A. Deiwan, "Getting the Most Out of Washing Deinking," *Tappi J.*, 65(7):64-68 (1982).

25. Healy, E. D., "Deinking Presents Profitable Alternatives: Two Different Technologies Discussed," *Paper Age*, (11): (1982).

26. Ghiz, R. E., "Deinking Advancement Toward Extended Raw Material Usage," *1982 TAPPl Papermaking Conference Proceedings*, Atlanta: TAPPI PRESS, 1982, pp. 263-268.

27. DeCeuster, J., and G. Papageorges "Physiochemical Aspects of Waste Paper Deinking by Washing," *Appita*, 35(2):(1981).

28. Casey, J. P., ed., *Pulp and Paper Chemistry and Chemical Technology*, 3rd edn., Vol. 1, Chap. 4.

29. Rangamannar, G., "Dispersion: An Effective Secondary Fiber Treatment Process for High Quality Deinked Pulp," *1989 TAPPI Pulping Conference Proceedings*, Atlanta: TAPPI PRESS, 1989, Book 1: 381-390.

30. Gilkey, M., H. Shinohara, and H. Yoshida, "Cold Dispersion Unit Boosts Deinking Efficiency at Japanese Tissue Mills," *Pulp & Paper*, 62(11):100-103 (1983).

31. Scarlett, T., "Printing Ink Formulations and Their Effect on Deinking," *1981 TAPPI Pulping Conference Proceedings*, Atlanta: TAPPI PRESS, 1981, Book 1: 181-183.

32. Burstall, M. L., "Inks and Deinking: Problems and Prospects," *Paper Technology and Industry*, (11): (1985).

33 Bassemir, R. W., "The Chemical Nature of Modern Printing Inks and Deinking," *1982 TAPPI Annual Meeting Proceedings*, Atlanta: TAPPI PRESS, 1982, Book 1: 99-101.

34. Williams, R. L., *Paper & Ink Relationships*, Roger Williams, Manhattan, Kansas, 1985.

35. Serchuk, A., "Radical Cure for Printing Ink Problems," *Chemical Business*, (12): (1985).

36. Nass, G., "The Economics of UV-Curing Inks," *American Ink Maker*, (6): (1975).

37. Hecht, M., "Cat's (Catalyzed Water-Based Coatings) Challenge UV Coatings on Sheetfed Press," *Paperboard Packaging*, 73(3):42, 44-46 (1988).

38. Visser, J. D., "Growth Areas in Ultraviolet/Electron Beam Markets," *Tappi J.*, 72(2):125-128 (1989).

39. Vanderhoff, J. W., "Deinking: The Industry's Position," *American Ink Maker*, (4): (1973).

40. Lyne, B. M., "Paper Requirements and the Evolution of Printing," *Tappi J.*, 72(5): (1989).

41. Hamilton, A. C., "Publication Flexo Uses Special Water-Based Inks," From an internal publication of General Printing Ink, Inc., Carlstadt, NJ.

42. Aspler, J. S., and Z. F. Perreault, "Newsprint Requirements for Water-Based Flexography, Part II: Newsprint Chemistry and Water-Based Inks," *Pulp & Paper Science*, 14(1):12-16 (1988).

43. Sun Chemical, "Task Force Formed to Develop an Effective Flotation Method for Newspapers Using Flexographic Inks," *Official Board Market*, (8): (1989).

44. Jarrehult, B., and R. G. Horacek, "Deinking of Waste Paper Containing Flexographic Inks," *1989 TAPPI Pulping Conference Proceedings*, Atlanta: TAPPI PRESS, 1989, Book 1: 391-405.

45. Seldin, I., "Xerographic Copy Recycling," *1985 TAPPI Pulping Conference Proceedings*, Atlanta: TAPPI PRESS, 1985, Book 2: 303-311.

46. Darlington, W. B., "A New Process for Deinking Electrostatically Printed Secondary Fiber," *Tappi J.* 72(1):35-38 (1989).

47. Sjostrom, E., *Wood Chemistry: Fundamentals and Applications*, Academic Press, 1981, pp. 172-174.

48. Rosen, M. J., *Surfactants and Interfacial Phenomena*, New York: Wiley-Interscience, 1978.

49. Jaycock, M. J., and G. A. Parfitt, *Chemistry of Interfaces*, New York: Wiley, 1981.

50. Bickerman, J. J., *Foams*, New York: Springer-Verlan, 1973.

51. Hornfeck, K., "Chemicals and Their Mode of Action in the Flotation Deinking Process," *Conservation and Recycling*, 10 (22): (1987).

52. Read, B. R., "Selection of Chemicals Within a Modern Deinking Plant," *Paper Technology and Industry*, (11): (1985).

53. Berrigter, A., "Development in Deinking," *World Pulp & Paper Tech.*, UK: Sterling Publ., 1990, pp. 88-90.

54. Anonymous, "Phosphates for Industry,"
From an internal publication of Monsanto
Industrial Chemical Co.

55. Falcone, J. S., and R. W. Spencer,
"Silicates Expand Role in Waste Treatment,
Bleaching, Deinking," *Pulp & Paper*, (12):
(1975).

56. Griffin, W. C., *J. Soc. Cosmetic
Chem.*, (1949).

57. Turai, L. L., and L. D. Williams, "The
Effect of HLB Factor of Nonionic Surfactants on
Deinking Efficiency," *Tappi J.*, 60(11):167-168
(1977).

58. Wood, D. L., "Alcohol Ethoxylates and
Other Nonionics as Surfactants in the Deinking of
Waste Paper," *1982 TAPPI Pulping Conference
Proceedings*, Atlanta: TAPPI PRESS, 1982,
Book 1: 435-446.

59. Suwala, D. W., "A Study of the
Deinking Efficiency of Nonionic Surfactants,"
1983 TAPPI Pulping Conference Proceedings,
Atlanta: TAPPI PRESS, 1983, Book 2: 533-541.

60. Anonymous, "Hydrophilic Polymers,"
From an internal publication of B. F. Goodrich.

61. Ing, F. J., and F. Carrion,
"Antiredeposition Agents," *Tensile Surfactants
Detergents*, 24(5): (1987).

62. Greminger, G. K., "Antiredeposition
Additives Face New Opportunities," *J. Am. Oil
Chemists Soc.*, 55(1): (1978).

63. Elsby, L., "Deinking of Waste Paper,"
From an internal publication of Berol Kemi AB,
1983.

64. Daneault, C., M. LaPointe, L.
Marchildon, "How to Improve the Brightness of
Pulp Obtained from the Deinking of Newsprint,"
1989 TAPPI Pulping Conference Proceedings,
Atlanta: TAPPI PRESS, 1989.

65. Korte, E. C., "Use of Chemicals in
Deinking," *Tappi J.*, (1978).

66. Kutney, G. U., "Hydrogen Peroxide:
Stabilization of Bleaching Liquors," *Pulp &
Paper Canada*, 86(12):182-189 (1985).

67. Dick, R. H., and D. H. Andrews, *Pulp
& Paper Canada*, 62(3):T201 (1965).

68. Colodette, J. L., S. Rochenberg, and C.
W. Dense, "Factors Affecting Hydrogen
Peroxide Stability in the Brightening of
Mechanical and Chemi-mechanical Pulps Part I:
Hydrogen Peroxide Stability in the Absence of
Stabilizing Systems," *J. Pulp and Paper Science*,
14(6):126-132 (1988).

69. Colodette, J. L., S. Rochenberg, and C.
W. Dense, "Factors Affecting Hydrogen
Peroxide Stability in the Brightening of
Mechanical and Chemi-mechanical Pulps Part IV:
The Effect of Transition Metals in Norway
Spruce TMP on Hydrogen Peroxide Stability," *J.
Pulp and Paper Science*, 15(3):79-83 (1989).

70. Colodette, J. L., S. Rochenberg, and C.
W. Dense, "Factors Affecting Hydrogen
Peroxide Stability in the Brightening of
Mechanical and Chemi-mechanical Pulps Part II:
Hydrogen Peroxide Stability in the Presence of
Sodium Silicate," *J. Pulp and Paper Science*,
15(1):3-10 (1989).

71. Fairbank, M. G., J. L. Colodette, T.
Ali, F. McLellan, and P. Whiting, "The Role of
Silicate in Peroxide Brightening of Mechanical
Pulp," *J. Pulp and Paper Science*, 15(4):132-135
(1989)

72. Burton, J. T., "An Investigation Into the
Roles of Sodium Silicate and Epsom Salt in
Hydrogen Peroxide Bleaching," *J. Pulp and
Paper Science*, 12(4): 95-99 (1986).

73. Colodette, J. L., S. Rochenberg, and C.
W. Dense, "Factors Affecting Hydrogen
Peroxide Stability in the Brightening of
Mechanical and Chemi-mechanical Pulps Part III:
Hydrogen Peroxide Stability in the Presence of
Magnesium and Combinations of Stabilizers," *J.
Pulp and Paper Science*, 15(2):45-51 (1989).

74 . Anonymous, *Hydrogen Peroxide for
Deinking Bulletin 133*, FMC Corp.

75. *Belloit Corporation Deinking Manual*,
2nd edn., 1979.

76. Altieri, A. M., and J. W. Wendell,
"Deinking," *TAPPI Monograph Series No. 31*,
Atlanta: TAPPI PRESS, 1967, Chap. 3.

77. Ellis, M. E., "How to Determine
Chemical Costs in the Bleaching of Mechanical
Pulps," *Pulp & Paper*, 61(6):129-132 (1987).

78. Ducey, M. "Chemicals Play Increasinq Role in Growth of Mechanical Pulping," *Pulp & Paper*, <u>61</u>(6): (1987).

79. Sellers, F. G., "New Hydrosulfite Route Reducers Groundwood Bleach Costs," *Pulp & Paper*, (1973).

80. Woodward, T. W., "Appropriate Chemical Additives Are Key to Improved Deinking Operations," *Pulp & Paper*, <u>60</u>(11):59-63 (1986).

81. Cruea, R. P., "Deinking: Laboratory Evaluations and Total System Concepts,"*Tappi J.*, <u>61</u>(6):27-30 (1978).

82. Mah, T., "Deinking of Waste Newspaper,"*Tappi J.*, <u>66</u>(10):81-83 (1983).

83. McCool, M. A., and C. J. Taylor, *Tappi J.*, <u>66</u>(8):69-71 (1983).

84. Giles, J. C., and J. M. Gliemmo, "Calcium Carbonate Scale Control in Secondary Fiber Deinking Plants," *1983 TAPPI Pulping Conference Proceedings*, Atlanta: TAPPI PRESS, 1983, Book 2: 509-513.

85. Wearing, J. T., and A. Garner, "Elimination of Thiosulfate From Paper Machine White Water Using Inorganic Oxidants," *Pulp & Paper Res. Inst. Canada*, (4): (1985).

Conversion Factors for SI Units

Quantity or Test	Value in Trade or Customary Unit	X	Conversion Factor	=	Value In SI Unit	Symbol
Area	square inches		6.45		square centimeters	cm^2
	square feet		0.0929		square meter	m^2
	square yards		0.836		square meter	m^2
	acres		0.405		hectares	ha
Basis Weight * or	lb (17x22-500)		3.760		grams per square meter	g/m^2
Substance	lb (24x36-500)		1.627		grams per square meter	g/m^2
(500-sheet ream)	lb (25x38-500)		1.480		grams per square meter	g/m^2
or Grammage* when	lb (25x40-500)		1.406		grams per square meter	g/m^2
expressed in g/m^2	pounds per 1000 sq ft (Paperboard)		4.882		grams per square meter	g/m^2
Breaking Length	meters		0.001		kilometers	km
Burst Index	$\dfrac{g/cm^2}{g/m^2}$		0.0981		$\dfrac{kPa}{g/m^2}$	
Bursting Strength	pounds per square inch		6.89		kilopascals	kPa
Caliper	mils		0.0254		millimeters	mm
Concora Crush	pounds		4.45		newtons	N
Edge Crush	pounds per inch		0.175		kilonewtons per meter	kN/m
Energy	British thermal units (Btu.)		1055		joules	J
Flat Crush	pounds per square inch		6.89		kilopascals	kPa
Force	kilograms		9.81		newtons	N
	pounds		4.45		newtons	N
Length	angstroms		0.1		nanometers	nm
	microns		1		micrometers	um
	mils		0.0254		millimeters	mm
	feet		0.305		meters	m
Mass	tons (2000 lbs)		0.907		metric tons	t
	pounds		0.454		kilograms	kg
	ounces (avd p)		28.3		grams	g
Mass per Unit Volume	ounces per gallon		7.49		kilograms per cubic meter	kg/m^3
	pounds per cubic foot		1.60		kilograms per cubic meter	kg/m^3
Puncture Resistance	foot pounds		1.36		joules	J
Ring Crush	pounds (for a 6 in. length)		0.0292		kilonewtons per meter	kN
Stiffness (Taber)	gram centimeters (Taber Units)		0.0981		millinewton meters	mN•m
Tear Strength	grams		9.81		millinewtons	mN
Tensile Breaking Load	pounds per inch		0.175		kilonewtons per meter	kN/m
	kilograms per 15 millimeters		0.654		Kilonewtons per meter	kN/m
Volume, Fluid	ounces (US Fluid)		29.6		milliliters	mL
	gallons		3.79		liters	L
Volume, Solid	cubic inches		16.4		cubic centimeters	cm^3
	cubic feet		0.0283		cubic meters	m^3
	cubic yards		0.765		cubic meters	m^3

*See TAPPI Technical Information Sheet 0800-01.

What is TAPPI?

Founded in 1915, TAPPI is the world's largest professional society of executives,
operating managers, engineers, scientists, and technologists serving the paper and related
industries. Total membership is approximately 32,000 with some 80% residing in the
United States. The remainder live in 76 other countries.

TAPPI is renowned for its industry publications. Members produce technical books,
reports, conference proceedings, course notes, home study courses, and videotapes
through TAPPI PRESS. *Tappi Journal*, distributed monthly to all members, is the leading
publication for technical information on the manufacture and use of pulp, paper, packaging,
and converted products. Through TAPPI, Association members develop, update, and
publish test methods and technical information sheets on which much of the industry
depends to analyze its products and processes.

TAPPI sponsors a variety of technical conferences, seminars, and short courses to foster
worldwide technical information exchange and enhance the professional development of
members.

For membership information, to order any of TAPPI's professional development products,
or to register for a meeting, please call TAPPI's toll-free service line:

1-800-332-8686 (U.S.)

1-800-446-9431 (Canada)

TAPPI's Vision

We are a global community of motivated individuals who lead the technical advancement of
the paper and related industries.

Together...

We provide outstanding educational and professional growth opportunities.

We serve as a worldwide forum to exchange technical information, promote research, and
recognize individual achievement.

We create success by the quality, timeliness, and innovativeness of our products and
services.

Integrity and fellowship characterize our association.

TAPPI PRESS Publications

Commercially Available Chemical Agents for Paper and Paperboard Manufacture
Fourth Edition, Revised
edited by John Farewell

Are you responsible for finding alternate suppliers for chemical additives? Or for improving properties of paper? This book provides the data to help you begin your search. It lists chemical agents that may be added at the wet end, size press or calendar stack to improve functional properties of paper.

Contains U.S. and European Products

The book is divided into U.S. (domestic) and European paper additives. Within each part is a listing of products by end use, a listing of products by producer, and supplier addresses. Within the listings by use, the products are arranged by chemical type and include physical form, charge, addition points, cure required, FDA status and supplier company name.

A project of the Chemical Processing Aids Subcommittee of the Papermaking Additives Committee, Paper and Board Manufacture Division.

Item Number: 01 01 R083
TAPPI Members: $20.00
List: $30.00

Dry Strength Additives
edited by Walter F. Reynolds

Features • Theory of Dry Strength Development • Surface Applications • Natural Gums as Dry Strength Additives • Cationic Starch as a Wet End Additive • Dry Strength from Wet Strength Agents • Acrylamide-based Polymers for Dry Strength Improvement of Paper • Dry Strength Additives as Practiced in Japan.

Item Number: 01 02 B044
TAPPI Members: $13.00
List: $20.00

Paper Machine Operations
Pulp and Paper Manufacture Series
Volume 7
edited by M.J. Kocurek and B.A. Thorp

Topics include • Paper Properties & Testing • Raw Materials • Headboxes • Wet End Chemistry • Pressing • Drying • Sizing • Reeling & Winding • Corrosion • Roll Vibration • Safety.

Item Number: 02 02 MS07
List: $120.00

Pulp and Paper Agitation: The History, Mechanics and Process
by D. Carl Yackel

Practical, useful information for design engineers and anyone involved in papermaking from the pulp mill through the dry end of the paper machine.

Includes a brief history of the development of agitators from 1800 to the present and detailed information to assist you in selecting agitators for your mill. The publication also presents calculations for safe shaft selection.

Item Number: 01 01 R177
TAPPI Members: $49.00
List: $74.00

1991 Focus 95+ Landmark Paper Recycling Symposium Proceedings
Atlanta, GA

Contains 25 papers, six question and answer sessions, and two wrap-up presentations from the symposium sponsored by API, IPST, NCASI, PIMA, TAPPI, and the USDA, Forest Service, Forest Products Laboratory. The papers, illustrated with over 200 figures and tables, cover • paper procurement • fiber balance • timber market implications • solid waste • tissue • newsprint • quality issues • printing and writing papers • paperboard • bleached packaging board • containerboard • technical challenges • process and product innovation • alternative uses of recycled fibers • environmental considerations.

Item Number: 01 05 1591
TAPPI Members: $79.00
List: $1 19.00

Recycling Paper: From Fiber to Finished Product
edited by Matthew J. Coleman

This two-volume anthology contains 150 papers on paper recycling presented at TAPPI conferences or published in *Tappi Journal* from 1981 to 1990.

Contains a variety of topics in three sections: Recycling Paper: A Critical Overview; The Processing of Secondary Fiber; and Paper and Board Manufacture Using Secondary Fiber.

Item Number: 01 01 R179
TAPPI Members: $96.00
List: $144.00

The Sizing of Paper
Second Edition
edited by Walter F. Reynolds

Addresses many dramatic technological advances. The 15 contributing authors, experts in their respective fields, emphasize correct procedures as well as problem solving and trouble-shooting. All those responsible for maximizing the performance of a sizing system will find this book useful. Contains seven chapters and 61 figures and tables.

Item Number: 01 02 B051
TAPPI Members: $46.00
List: $68.00

Call TAPPI toll-free to order from TAPPI PRESS
1-800-332-8686 (U.S.) • 1-800-446-9431 (Canada)

Please allow 4-6 weeks for delivery. This offer and prices are subject to change without notice, and availability of inventory.